Nonlinear Dynamics and Fractals New Numerical Techniques for Sedimentary Data

Gerard V. Middleton
McMaster University, Hamilton ON

Roy E. Plotnick
University of Illinois, Chicago IL

David M. Rubin
U.S. Geological Survey, Menlo Park CA

1926 **SEPM**
Society for
Sedimentary Geology

SEPM Short Course No. 36

ii

These SEPM Short Course Notes have received independent peer review. In order to facilitate rapid publication, these notes have not been subject to the more stringent editorial review required for SEPM Special Publications.

ISBN 1-56576-021-2

Additional copies of this publication may be ordered from SEPM. Send your order to

SEPM
1731 E. 71st Street
Tulsa, Oklahoma 74136–5108
U.S.A.
Tel: 918–493–3361   Fax: 918–493–2093

Contents

Preface

Our intention in producing these notes has been to provide sedimentary geologists with an introduction to the new techniques for analysing experimental and observational data provided by the rapid development of those disciplines generally known as Fractals and Nonlinear Dynamics ("chaos theory").

We give a general introduction to a minimum of theory, but devote most of our space to show how these ideas are useful for interpreting sedimentary data. It seems to us that the main applications are likely to be to time series or spatial profiles or two-dimensional maps or images. Sedimentary geologists deal every day with actual time series, such as measurements of current velocity or suspended concentration at a station, or with "virtual" time series, such as stratigraphic sections, well logs, or topographic profiles—yet few geologists know much about the new numerical techniques available to analyse such data.

Much of the primary literature is published in journals or books that sedimentary geologists do not read, and often in language that they cannot understand. Though we have no intention of avoiding mathematics completely (after all, the techniques to be described *are* numerical techniques), we hope that these notes are written in a language that most geologists can readily understand, and make use of examples relevant to our discipline. An extensive **glossary** is provided as Appendix I, which should make it easier for those who wish to explore the primary literature (much of it published in math or physics journals).

We thank David Goodings for critical reading of earlier drafts of parts of these Notes, Dave Cacchione and Ralph Hunter (both at U.S. Geological Survey, Menlo Park) for reviewing Chapter 5, and Henry Abarbanel for his comments on almost the whole MS. Bruce Malamud provided recent references on rescaled range analysis. Martin Perlmutter and Texaco, Inc., provided the well data used in Chapter 1. Henry Abarbanel and Upmanu Lall provided preprints of papers about the Great Salt Lake, and gave us permission to reproduce material from papers in press.

Roy Plotnick's work has been funded by grants from the National Science Foundation to Plotnick and Prestegaard (EAR-890484), and to

Plotnick from Texaco, Inc. Gerry Middleton's work has been funded by the National Science and Engineering Council of Canada.

Dave Rubin thanks George Sugihara (Scripps Institute of Oceanography) and James Theiler (Los Alamos National Laboratory) for enjoyable discussions. Dave Rubin's work was funded in part by NASA (Grant No. W-17,975).

Finally, Gerry Middleton set these notes in type, using LaTeX. This, and TeX, from which it is derived is a type setting program produced by Donald Knuth and Leslie Lamport, who donated the copyright to the American Mathematical Society. Several free versions of this program are available—the one Gerry used is called emTeX, and was written by Eberhard Mattes. TeX is now used by many math and physics journals, including the AGU journals, and deserves to be more widely adopted by geologists. We owe a special debt to its authors, who donated their time freely to the scholarly professions, so that the costs of typesetting might be considerably reduced.

Gerry Middleton (`middleto@mcmaster.ca`)
Roy Plotnick (`plotnick@uic.edu`)
Dave Rubin (`dr@octopus.wr.usgs.gov`)

Chapter 1

INTRODUCTION TO FRACTALS

Roy E. Plotnick
Department of Geological Sciences
University of Illinois
Chicago IL 60680 USA

So, Nat'ralists observe, a Flea
Hath smaller fleas that on him prey,
And these have smaller yet to bite 'em
And so proceed ad infinitum.

Jonathan Swift (1733)

1.1 Introduction

The incorporation of fractal concepts into the natural sciences, especially physics, has been swift. This growth has also occurred, albeit at a much slower pace, in the earth sciences, including geomorphology, sedimentology, stratigraphy, and petroleum geology (Korvin, 1992; Turcotte, 1992, 1994a; Barton and LaPointe 1995a,b). Nevertheless, many earth scientists remain unfamiliar with fractal concepts and applications and are perhaps even suspicious that it is a fad. In this chapter we will demonstrate that fractal models and methods allow the geologist to quantify many concepts that have long been intuitive, while also providing new and fruitful ways of looking at data. In addition, as we will show in later chapters, fractals are important in the interpretation of chaos and non-linear dynamics. We believe that fractal meth-ods will eventually become part of the standard toolkit of any quantitatively oriented geologist, to the same extent that calculus or statistics are currently.

1.2 Why fractals, or why artists, have been necessary.

The standard tools of a draftperson include a t-square, triangle, french curve, and protractor. Using these items, it is possible to draw nearly all of the myriad objects that comprise the manmade world. When we now try to draw trees, clouds, brachiopods, or river drainage networks, however, these tools must be discarded. Instead, the abilities of an artist to portray the irregularity of nature must be called upon. As stated by Mandelbrot (1983): "Clouds are not spheres, mountains are not cones, coastlines are not circles, and bark is not smooth, nor does lightning travel in a straight line." Standard Euclidean geometry, while ideally suited to the description of manmade objects (Figure 1.1), is generally inadequate to describe natural objects. For this reason, the fractal geometry invented by Benoit Mandelbrot, which allows the realistic description and simulation of natural shapes (including the "shape" of time series) has received acceptance in both the scientific and artistic communities.

At the same time, fractals have become a standard technique among motion picture spe-

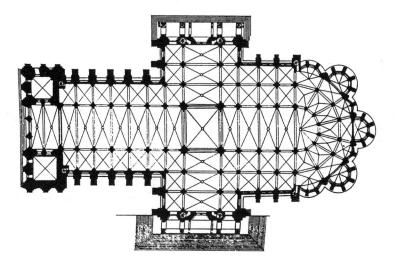

Figure 1.1: Floor plan, Chartres cathedral.

cial effect artists for the generation of real-istic artificial planets and landscapes. Simi-larly, fractal oriented computer scientists have teamed with biologists to produce model or-ganisms, such as plants and corals, of startling realism (Prusinkiewicz and Lindenmayer, 1990; Kaandorp, 1994). The fidelity of such geometric constructs to real objects strongly suggest that fractals can be used not just to describe, but to model.

1.3   What are fractals? Self-similarity and the fractal dimension

No generally agreed upon definition of a fractal exists (Feder, 1988); nevertheless, any definition includes the concepts of both *self-similarity* and *fractal dimension*. To understand the concept of self-similarity, it is useful to consider two famil-iar Euclidean geometric objects; a line segment and a circle (Figure 1.2). If we take the line seg-ment and put it underneath a magnifying glass, the object we see is still a line segment. If we

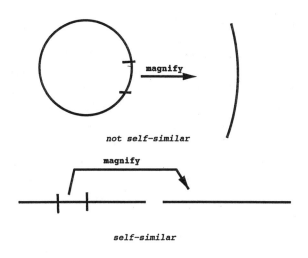

Figure 1.2: Self-similarity in Euclidian objects

increase the magnification, we still see a line
segment. In other words, any part of a line segment looks exactly like the entire line segment;
line segments are thus self-similar.

The circle, in contrast, is not self-similar. If
we magnify a portion of the circle, we first see an
arc, and then lines of progressively less apparent
curvature. It is this property that allows us to
take a tangent of the circle.

A self-similar fractal object, the Koch curve,
is illustrated in Figure 1.3D. The self-similarity
of the Koch curve can be understood by examining its construction (Figure 1.3A–D; see Feder,
1988). The procedure begins with the simple
curve shown in Figure 1.3A. Each linear segment in now replaced with a one-third size du-
plicate of the original curve (Figure 1.3B). This
procedure is then repeated (Figure 1.3C) until
the final curve (Figure 1.3D) is produced. Several things to note:

1. the final curve is produced by an iterative
   operation, rather than a formula; this is
   characteristic of fractal objects (it also, as
   will be seen in Chapter 2, is why fractals
   are necessary to describe some iterative dynamic models);

2. the iterative operation for this ideal geometric form is repeated endlessly, so that
   it is self-similar over all possible scales of
   observation;

3. as a result, the curve *never* becomes
   straight, no matter how small the scale; i.e.,
   it is non-differentiable.

If a line segment and a Koch curve are both
self-similar, what makes the latter a fractal? To
understand this, we need to compare the concepts of topological, Euclidean, and fractal dimension. Examine the various curves shown
in Figure 1.4; you may want to think of them
as river planforms. The paper surface represents a two-dimensional Euclidean space which
the curves occupy. Now imagine a boat sailing

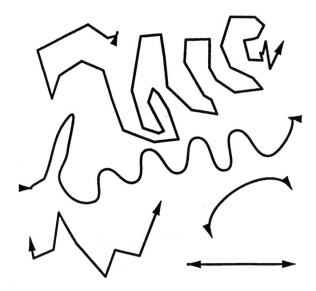

Figure 1.4: Topological dimension: All of these
curves have a topological dimension of 1, but
have different fractal (Hausdorff–Besicovitch)
dimensions.

on one of these rivers that is the same width
as the river. As a result, the boat can only
move forward or back; i.e, to the boat, the
line is one-dimensional, no matter how curved
it is. Another way to think of this is that if
the curves were made of some flexible material,
they could be pulled to form a straight line. The
same is true of the Koch curve is Figure 1.3.
All of these lines have a topological dimension
(D_T) of 1. In terms of topological dimension,
a point (or set of points) has a dimension of 0,
any non-intersecting line has $D_T = 1$, any non-
intersecting surface has $D_T = 2$, and so on.

Despite all of the lines in Figures 1.3 and 1.4
having the same topological dimension, they are
clearly different in shape. In particular, the
more intricate the line, the more "densely" it
covers the surface of the paper; i.e., the more
completely it occupies the two-dimensional Euclidean space. One can, perhaps, imagine a line
which would be so intricate that it would totally
cover the page. Such a line, although it would
still be topologically 1-dimensional, actually oc-

A

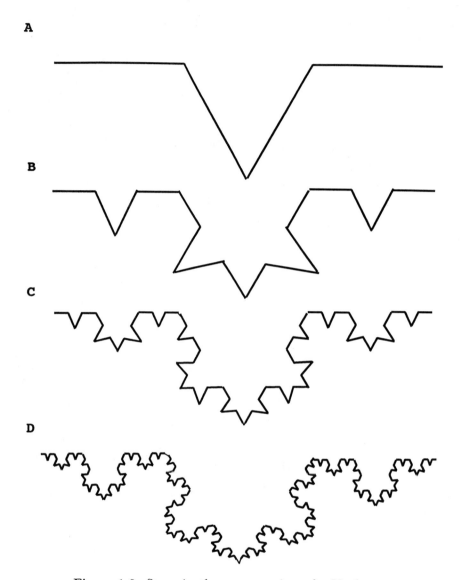

B

C

D

Figure 1.3: Steps in the construction of a Koch curve

cupies a two-dimensional Euclidean space. It is this extent of filling by an object of a lower topological dimension of a space of higher Euclidean dimension that the fractal dimension attempts to measure.

How is the fractal dimension determined? Once again look at the standard Euclidean line segment (Figure 1.5A) and ask: how long is it? Imagine, for the moment, that you are given an unruled 1 m long stick (r = stick length) and are asked to measure the length (L) of the line segment. Working swiftly, you find that the line is 1 stick length long (N = number of stick lengths). After hours at a workstation, you determine that L must be $r \times N = 1$ m long. Just to be sure, you repeat the process using a stick $\frac{1}{3}$ of a meter long and find that your line segment is 3 stick lengths long; L again equals 1 m. Having nothing better to do, you repeat with a $\frac{1}{9}$ m long stick and get the same result. In fact, no matter what size stick you use, you always end up with a value of 1 m. The line segment can be said to have a characteristic length.

You now receive a grant to repeat the experiment with the Koch curve (Figure 1.6). Starting initially with the 1 meter stick, you find that the curve is 1 stick lengths or 1 meters long. Now you use the $\frac{1}{3}$ m stick and find the curve is 4 stick lengths long. Multiplying, you obtain a curve length of $1\frac{1}{3}$ m! Agitated now, you repeat the experiment with a $\frac{1}{9}$ meter stick, yielding an L of $1\frac{7}{9}$ m! In fact, the smaller the measuring stick, the larger the measured length. The Koch curve thus lacks a characteristic length; its size is dependent on the scale of the measuring device. The lack of a characteristic length, along with it self similarity, is what makes a Koch curve a fractal.

Another way of looking at the same situation is that a line segment of unit length can be viewed as consisting of a series of smaller line segments, e.g., of 3 segments, each $\frac{1}{3}$ long. Each of these line segments can be considered, due to self-similarity, to be identical to the original line segment scaled down by a factor $r = \frac{1}{3}$. This can then be repeated, with each smaller segment scaled down by an additional $\frac{1}{3}$, yielding 9 segments of length $\frac{1}{9}$. The size of the whole object is always $N \times r = 1$, no matter what value of r is used.

Now consider a square 1 unit on a side (Figure 1.6B). We can divide it into $N = 9$ identical parts, each with a linear dimension $\frac{1}{3}$ ($r = \frac{1}{3}$) of the original. In this case, $N \times r^2 = 1$, no matter what r is chosen.

We can generalize these relationships (Voss, 1988) to:

$$Nr^D = 1 \qquad (1.1)$$

where N is the number of parts an object of unit size is divided into, r is the ratio of size of the parts to that of the whole object, and D is the *fractal dimension*. This equation can be easily solved for D:

$$D = \log N / \log(1/r). \qquad (1.2)$$

You should immediately note that for the line segment $D = 1$ and for the square $D = 2$; for a cube D would equal 3. For these Euclidean objects, the fractal dimension and the topological dimension are equal.

For the Koch curve the situation is different. At each iteration, the size of the pieces are reduced by $1/3$ ($r = 1/3$). However, each piece is divided into four parts ($N = 4$). As a result, $D = \log(4)/\log(3) = 1.2618\ldots$ The fractal dimension of the Koch curve thus exceeds its topological dimension, reflecting the ability of the Koch curve to cover a 2-dimensional surface. This is the second defining characteristic of a fractal object. Notice also that the fractal dimension is not integer, although this is not a necessary condition for an object to be fractal.

The reason for the lack of a characteristic size of the Koch curve now becomes clear. As shown above, $L = N \times r$. Since (from Equation 1.1) $N = 1/r^D$,

$$L = 1/r^{D-1}. \qquad (1.3)$$

Since $D > 1$ for the Koch curve, L increases as r decreases. Note that this relationship can be

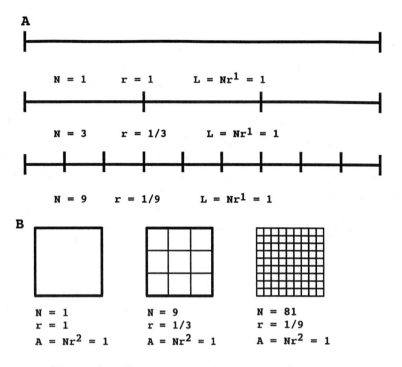

Figure 1.5: Fractal dimension of Euclidean objects.

rewritten:

$$\log L = (1 - D) \log r, \qquad (1.4)$$

i.e., $(1 - D)$ is the slope of a plot of $\log L$ versus $\log r$. For objects of non-unit length,

$$L \propto 1/r^{D-1} \qquad (1.5)$$

and the same relationship holds.

The form of the fractal relationship, in which the measured size of an object is a power function of the measurement scale, is a familiar one to geologists. Numerous well-established empirical relationships, such as the Gutenberg–Richter law of earthquake frequencies or Hack's law relating the length of the longest river in a drainage basin to the area of that basin (Turcotte, 1992; Korvin, 1992) are power functions. In addition, values of the exponent in these relationships are frequently non-integer. As will become clear, this power law relationship is a direct consequence of the self-similar or fractal structure of the pattern being investigated.

To summarize to this point:

1. all geometric objects, whether they are fractal or not, have a fractal dimension;

2. in order for an object to be considered fractal, it must be both self-similar and have a fractal dimension greater than its topological dimension;

3. fractal objects lack a characteristic scale;

4. fractal patterns are described by power law relationships, identical to those frequently encountered in geology.

1.4 Are real things fractal? The fractal dimension of natural objects

It must be pointed out that fractals are geometric models, just as circles, squares, and cubes are. Real objects are not fractals, in the sense that the earth is not really a sphere. Instead, many real objects have a fractal structure; i.e., they can be mathematically modelled

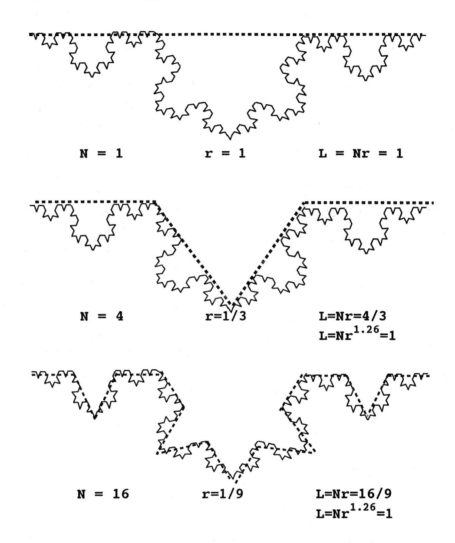

Figure 1.6: Fractal dimension of a Koch curve.

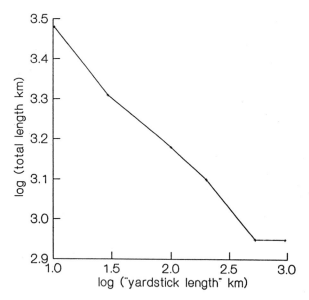

Figure 1.7: Richardson plot, British coast.

using fractals. In addition, unlike the Koch curve and other fractal objects, real objects are fractal only over some range of scales. As pointed out by Peitgen and Richter (1986, p.5) "...no structure can be magnified an infinite number of times and still look the same. The principle of self-similarity is nonetheless realized approximately in nature ..."

The analysis of the Koch curves shows one approach to determining if a real object, such as a river planform or a coastline is fractal. The technique, which actually has a long pedigree (Mandelbrot, 1983), involves determining how many linear units of a certain length are needed to span an object. This procedure is repeated for numerous sizes of the linear unit. The total length measured is then plotted, on a log-log graph, against the size of the measuring unit, producing what is termed a "Richardson plot," after its inventor (Mandelbrot, 1983); and the resulting estimate is the "divider dimension". If the object has a fractal structure, the result should be a line with a slope equal to $1 - D$.

The classic example of the use of the Richardson plot method is the analysis of the length of the coastline of Britain (Mandelbrot, 1983;

Carr and Benzer, 1991). Figure 1.7 shows the Richardson plot for coastline length versus size of the measuring "yardstick;" the slope of the least-squares fit is -0.27, yielding a fractal dimension D of 1.27. Notice that this value is nearly identical to the Koch curve value of 1.26, suggesting how the Koch curve can be used as a "model" of the coastline. Carr and Benzer (1991) analyzed seven additional coastlines and found estimates of D ranging from 1.02 (East shore, Gulf of California) to 1.19 (North coast of Australia).

An excellent example of the application of this approach to a geomorphic situation can be found in Snow (1989), who examined a set of 12 stream channel planforms and compared their Richardson plots with those of idealized planforms. He obtained fractal dimensions for the actual streams in the range of 1.04–1.38, whereas the artificial planforms failed to show a fractal structure. As discussed above, the empirical plots are linear over a limited range of values, which is typical of "real" fractal plots.

This method has also been fruitfully applied to the analysis of ammonoid sutures (Garcia-Ruiz, Checa, and Rivas, 1990; Boyajian and Lutz, 1992). Boyajian and Lutz used the fractal dimension as a metric for the measure of morphological complexity. Simple sutures, similar to those found in nautiloids, have a fractal dimension close to 1. More complex sutures can have fractal dimensions of up to 1.64. In general, they found that the range of sutural complexity, as measured by the fractal dimension, increased during evolution of the group.

An alternative method for estimating the fractal dimension is the "box dimension" (Voss 1986, Feder 1988). The object of interest is covered by non-overlapping boxes of linear dimension l. The number of boxes N containing at least part of the object are then counted. The number of boxes is related to l and D by

$$N \propto l^{-D}. \tag{1.6}$$

This method is illustrated in Figures 1.8 and

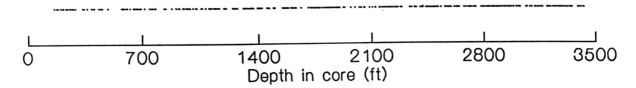

Figure 1.8: Location of gamma ray peaks in a core.

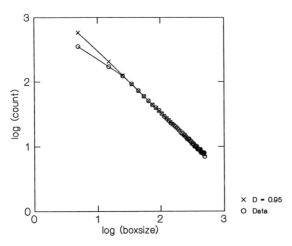

Figure 1.9: Box count of data in Figure 1.8.

1.9. Figure 1.8 is the depth distribution of gamma-ray emission peaks ("kicks") in a 3435 ft long well log from the Triassic Taylorsville basin of North Carolina. Peaks were identified by applying a cutoff value to the log; values above the cutoff were assigned values of 1, those below 0. Figure 1.9 is the box count plot of this data set; the data are very well fit by a regression line corresponding to a fractal (box) counting dimension of 0.95. The high value of D is a direct consequence of the large number of values identified as peaks (831 out of 3435).

A similar method was used by Chopard, Herrmann, and Vicsek (1991) to analyze the structure of mineral dendrites. For dendrites on lime-stone surfaces from Germany (Solnhofen) and Greece they found fractal dimensions of 1.78. Dendritic patterns of similar fractal dimension have been produced using diffusion limited aggregation (DLA) models (see Appendix I: Aggregation).

Fractal methods have also been used to characterize porosity in sedimentary rocks (Krohn, 1988a,b; Wong, 1988; Thompson, 1991). Using SEM images of sediment pores at different magnifications, Krohn identified surface features (grain edges, bumps, pits) based on changes of contrast. She discovered that the pore spaces of nearly all sedimentary rocks studied (the exception was a novaculite) showed fractal structure of some range of magnitude, with fractal dimensions in the range of 2.3–2.9. Wong (1988) found a similar range of fractal dimensions using neutron scattering.

1.5 Some useful fractal models

The Koch snowflake curve is probably the most familiar fractal model. Nearly as familiar is the Cantor dust (Mandelbrot, 1983; Plotnick, 1986; Korvin, 1992). A Cantor dust is easily generated (Figure 1.10). Begin with a line segment of unit length;. the middle third of the line segment is now erased, leaving two lines segments, each 1/3 the length of the original. The process

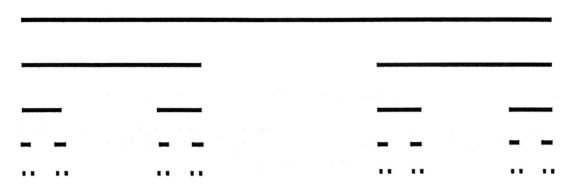

Figure 1.10: Generating a Cantor dust

is now repeated with the smaller line segments, producing 4 line segments, each 1/9 the length of the original. There is one gap of length 1/3 and two gaps of length 1/9 between them. The process is repeated indefinitely, leaving a set of points (the Cantor dust) separated by gaps of a wide range of lengths. The process is self-similar, so that the Cantor dust is a fractal set. Since two pieces, each 1/3 the length of the previous piece are produced at each iteration, the fractal dimension $D = \log 2/\log 3 \approx 0.6309$.

Alternative versions of the Cantor dust can be produced by changing the size of the gaps (Mandelbrot, 1983; Plotnick, 1983). A randomized version of the dust, known as a Lévy dust can also be produced (Figure 1.11). For both Lévy and Cantor dusts, the distribution of gaps follows the "uniformly self-similar law of probability" (Mandelbrot, 1965). The cumulative frequency distribution U_ϵ of gap lengths, greater or equal to length u, is described by:

$$P(U_\epsilon \geq u) = \left(\frac{u}{\epsilon}\right)^{-D}, \qquad (1.7)$$

where ϵ is the minimum level of resolution and D is the fractal dimension.

One possible use of the uniformly self-similar law of probability is in the estimation of confidence intervals on biostratigraphic ranges. That is, given the empirical range of a fossil organism, what are the probabilities of additional finds beyond the range? A model proposed by

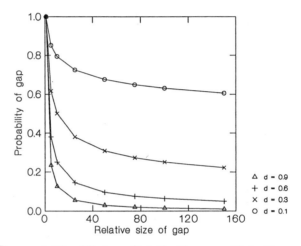

Figure 1.12: Hiatus distributions; self-similar probability law.

Strauss and Sadler (1989) and discussed in detail by Marshall (1991) assumes that fossil horizons are randomly distributed within the known range. The confidence intervals on first or last occurences are then a function of the total number of occurrences within the range.

Some theoretical biostratigraphic distributions are shown in Figure 1.11. All have the same total range and number of horizons. They range from highly regular, to random, to fractal (a Lévy dust), to highly clumped. Under the random model, all of these sequences would have the same confidence intervals. Yet it is intuitive that the more clumped the sequence the

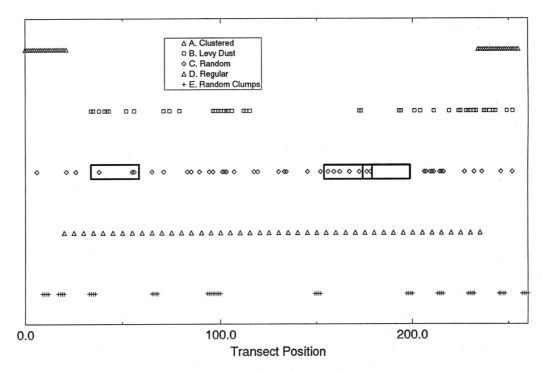

Figure 1.11: Synthetic stratigraphic sections or transects.

longer the confidence intervals should be; i.e., not only the the number of horizons matters but also their distribution. This is reflected in the uniformly self-similar law, which predicts much longer confidence intervals than corresponding random prediction. Figure 1.12 graphically illustrates the law for values of D from 0.1 to 0.9; note the very high probability of large gaps associated with low D values.

Plotnick (1986; Plotnick and Prestegaard, 1993) used the Cantor dust as a model for the temporal distribution of stratigraphic hiatuses; i.e., diastems and unconformities. The model is based on a modification of the Cantor dust concept known as the Cantor bar. A stratigraphic section is viewed as being composed of physical rock units that were deposited in numerous small increments of time; i.e., each unit of time has a particular thickness of rock associated with it. Technically, the rock units are known as the *mass* and the set of time intervals is the *support* of the mass (Mandelbrot, 1989).

For example, assume that a section consists of 1,000 m of sediment ($M = 1000$) that was deposited over a total time interval of one million years ($L = 10^6$); the total rate of deposition $R = 10^{-3}$ m/yr.

Now assume that the distribution of time intervals of deposition follows the uniformly self-similar law of probability and that equal increments of sediment are deposited at each time interval. For example, at the second stage of the generation of the Cantor bar (Figure 1.10) there would be four intervals of sedimentation, each 111,111 years long, each comprising 250 m of rock. There would be 3 hiatuses, one 333,333 years long and two 111,111 years long. The next generation (Figure 1.10) would add 4 hiatuses, each 37,037 years long, with each interval of sedimention being 37,037 yrs long and comprising 125 m of rock. Note that whereas the time intervals of sedimentation are getting shorter, the total amount of rock is conserved. The relationship between the fraction l of total time L represented by a single time interval to the fraction m of total thickness M deposited during this interval is

$$m = l^{\alpha}, \qquad (1.8)$$

where $\alpha = 0.6039\ldots$ and is known as the Lipschitz-Hölder exponent (Feder, 1988). The concept of the Lipschitz-Hölder exponent will be critical in our later discussion of multifractals.

The rate r of deposition (i.e., "density") for each time interval can be determined from

$$r = m/l = Rl^{\alpha-1}. \qquad (1.9)$$

Notice that as l decreases, r increases. This model thus produces results that are consistent with the empirical observations of Schindel (1980) and Sadler (1981), who demonstrated that there is a negative log-linear relationship between measured rates of sedimentation and the time period over which they are measured (see also Sadler and Strauss, 1990). In addition, numerical experiments by Korvin (1992) and Thorne (1995), using stratigraphic and sedimentary process models, support the concept that stratigraphic units and their corresponding hiatuses are fractal.

Cantor dusts lie along a line; similar objects can occupy a plane (a Sierpinski carpet) or a volume (Menger sponge). These objects have void spaces on all scales. They thus have been used as models for porous media, such as sedimentary rocks or soils (Wong, 1988; Turcotte, 1992).

1.6 Multifractals: When One Dimension Isn't Enough

The Cantor bar model introduces the concept that geologists are often concerned not just with the presence or absence of an item (fossils, sand beds) at a particular location or time but with how much of the item (number of fossils; sand percentage) is present. For example, the Cantor bar model could describe the occurence of fossils

in a stratigraphic sequence, with the locations of the intervals representing fossiliferous horizons and the mass at each horizon representing the number of fossils present.

We can characterize this patterns using the Lipschitz-Hölder exponent α. Equation 1.8 can be solved for α:

$$\alpha = \log m / \log l, \qquad (1.10)$$

where m is the fraction of the total material (or mass) in a particular unit and l is the fraction of the total thickness of the sequence represented by the unit. Notice that the use of α lets sequences of different lengths and total masses to be compared. Any series of abundance variations can thus be transformed into a series of Lipschitz-Hölder exponents. In the case of the Cantor bar, each fossiliferous horizon has the same value of α and the distribution of these horizons is described by the single fractal dimension $D = 0.6039$ (which in this case is identical to α).

There is an obvious problem with this model, however; the same number of fossils is present at each horizon! In actuality, of course, the number of fossils at each horizon could differ markedly. In addition, it is also possible that the fossils are not concentrated in individual horizons separated by gaps, but are present continuously throughout the sequence, with some horizons having higher concentrations than others. Consequently, in the case of real data the values of α will differ greatly between different horizons. A unit with abundant fossils will have a much higher value of α than a unit of equal thickness with few fossils.

Figure 1.13 illustrates a synthetic fossil abundance sequence with wide extremes of abundance. This sequence is produced by:

1. splitting the sequence in half and randomly assigning a fraction p of the fossils to one half and a fraction $(1-p)$ to the other half; in this example p is a Gaussian random variable with a mean of 0.6;

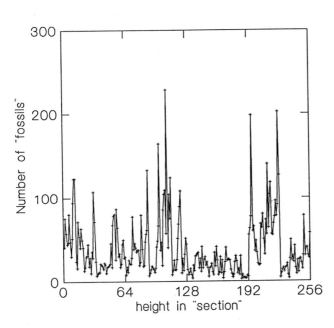

Figure 1.13: Synthetic abundance sequence produced by random binomial multiplicative process.

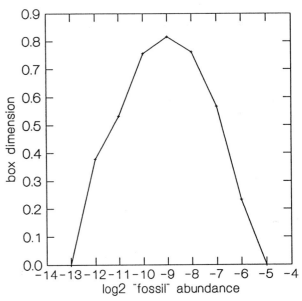

Figure 1.14: Multifractal spectrum (box counting dimension) for series in Figure 1.13.

2. splitting each half-sequence in half again and once more assigning a fraction p to one half and $(1 - p)$ to the other half.

This process is continued to produce the desired number of subsequences. The sequence in Figure 1.13 represents 8 generations and thus 256 "beds", with a total of 10,000 fossils. This self-similar procedure is a random version of the binomial multiplicative process (Feder, 1988; Mandelbrot, 1989; Plotnick and Prestegaard, 1995).

An alternative view of this model is that it generates the probability that a fossil will be found in a particular part of the sequence. Although the probability in some horizons may be very low, it is never zero.

Consider a particular value, or range of values, of fossil abundance in the sequence. This value can occur never, rarely, or frequently. For example, in Figure 1.13 very high or very low values occur infrequently; intermediate values are much more common. The spatial distribution of this value can be described by its fractal dimension. Now consider a different abundance value; its distribution will be described by a different fractal dimension. Each abundance value (and its corresponding α) will have its own fractal dimension associated with it; i.e., the distribution of material cannot be completely specified by a single fractal dimension. These sequences are thus *multifractal*.

This is illustrated in Figure 1.14. Each point represents the fractal (box) dimension of the occurence of a particular range of fossil relative abundances. Intermediate abundances are more common, so they have a higher value of D, while very high or low abundances are rare and thus correpond to lower D values. The resulting plot is a *multifractal spectrum*.

An alternative view of this concept is to consider the probabilities of a particular value of α in a sequence; some values occur with a higher probability and others with a lower probability. One approach to treating multifractal distributions (Feder, 1988; Mandelbrot, 1989) is to:

1. determine the probability distributon of the α values;

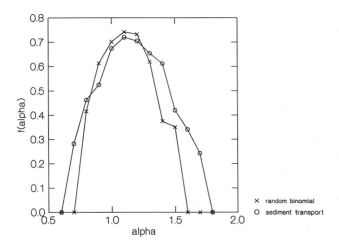

Figure 1.16: $\alpha/f(\alpha)$ spectrum for series in Figures 1.13 and 1.15.

2. for each value of α, calculate the value $f(\alpha)$ by the formula

$$f() = 1 - \log(\text{probability of } \alpha)/\log(l);$$
(1.11)

3. plot $f(\alpha)$ versus α.

$f(\alpha)$ is thus a transformation of the probability distribution of the standardized abundances.

Figure 1.16 gives the $f(\alpha)$ versus α curve for the synthetic data in Figure 14. The shape is similar to that obtained through box counting (Figure 1.14); i.e, the values of $f(\alpha)$ are, to a first approximation, estimates of the fractal dimension of the set of points of value α (Mandelbrot, 1989).

An example of a sequence with a multifractal structure is the 5 hour time series of sediment transport shown in Figure 1.15. The data set comprises 302 consecutive 1 minute samples of transported bedload in the East Rosebud River, Montana (see Prestegaard and Plotnick, 1995 for details).

Figure 1.16 plots the α versus $f(\alpha)$ for this data set. Note the similarity between the shape of the empirical multifractal curve and that produced by the randomized binomial multiplicative process.

What might produce a multifractal distribution? Stanley (1991) suggested that they may characterize systems where the underlying dynamics are governed by random multiplicative processes, such as those that produce the random binomial sequence described above. For example, the occurence of fossils is governed by a large number of biologic and taphonomic processes which act over a wide range of spatial and temporal scales. The occurrence of a large number of fossils at a horizon can only occur if the probabilities of all controlling factors are simultaneously high (e.g., P(fossil preservation)= p(large original population numbers) × p(rapid burial) × p(noncorrosive poor waters)...). Similarly, sediment transport fluctuations may result from multiplicative and non-linear interactions among the heterogeneous material on the stream bed, the material in transport, and local fluctuations in stream hydraulics (Prestegaard and Plotnick, 1995).

It should be pointed out that multifractals are a far more complex and subtle subject than portrayed here; interested readers are referred to Feder (1988), Mandelbrot (1989), and papers in Bunde and Havlin (1991).

1.7 Lacunarity: Beyond the Fractal Dimension

Although fractal methods are starting to become part of the standard approach to the analysis of spatial patterns, they are often inadequate to describe the full range of real patterns. Real patterns may or may not be fractal; when fractal structure exists, it may be only over a limited range of orders of magnitude; and patterns with the same fractal dimension may still look different; i.e., have different "textures."

This last point is illustrated in Figure 1.17, which illustrates two different forms of the Cantor set. In the first upper example, N=2 and r=1/4, so that D = 0.5. In the lower example,

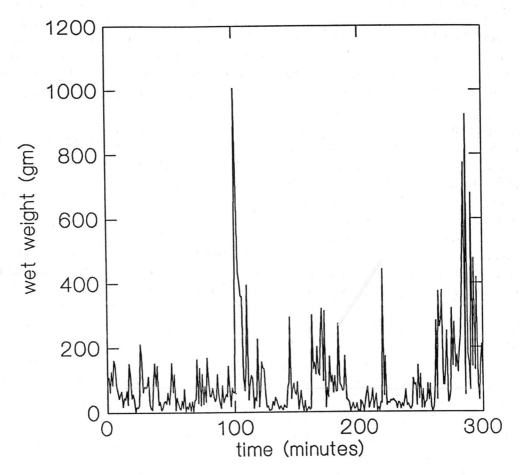

Figure 1.15: Five hour sequence of sediment tranport (Prestegaard and Plotnick, 1995).

Figure 1.17: Cantor sets, both with $D = 0.5$.

$N = 3$ and $r = 1/9$, so D again equals 0.5. Despite having the same fractal dimension the two patterns clearly look different; the upper set appears "gappier." Mandelbrot (1983) termed this difference in appearance of sets with the same fractal dimension "lacunarity," from *lacuna*; the upper set has higher lacunarity than the lower set.

Examine again the synthetic fossil sequences shown in Figure 1.11. All of the sequences have the same length ($M = 256$) and the same number of fossiliferous horizons ($S = 44$). The percent P of "beds" containing fossils in thus 0.172. Sequence A's horizons are all clustered at the extremes of the line; sequence B's approximate a fractal Lévy dust; sequence C's are randomly placed; and sequence D's are regularly distributed. In sequence E, the horizons occur in clumps of 4 units thick but these clumps are themselves randomly distributed: i.e., there are two distinct scales to the pattern.

Sequence D has a distinct characteristic; when examined at any scale greater than the spacing between the horizons, one part of the sequence looks identical to any other part of the sequence. In other words, the appearance of the sequence does not depend on position; the sequence is translationally invariant. This is in sharp contrast to sequence A, where the ends of the sequence look quite different from the middle part, so that the sequence is not translationally invariant. Notice that the more clumped a sequence is, and thus the larger the relative size of the gaps, the less translationally invariant it is. In a general sense, therefore, the greater the lacunarity of an object, the less translationally invariant it is (Gefen et al., 1983).

In order to quantify lacunarity, we will use the "gliding box" algorithm of Allain and Cloitre (1991; see also Plotnick et al., 1993). A box of length r is placed at the origin of one of the sequences (Figure 1.11). The number of fossiliferous horizons within the box (box mass = s) is now determined. The box is moved one unit along the sequence and the box mass is again

counted. This process is repeated over the entire sequence, producing a frequency distribution of the box masses $n(s, r)$. This frequency distribution is converted into a probability distribution $Q(s, r)$ by dividing by the total number of boxes $N(r)$ of size r. The first and second moments of this distribution are now determined:

$$Z(1) = \sum SQ(s, r) \qquad (1.12)$$
$$Z(2) = \sum S^2 Q(s, r). \qquad (1.13)$$

The lacunarity for this box size is now defined as:

$$\Lambda(r) = Z(2)/(Z(1))^2. \qquad (1.14)$$

This calculation is repeated over a range of box sizes, ranging from $r = 1$ to some fraction of M (usually $M/2$). A log-log plot of the lacunarity versus the size of the gliding box is then produced. The lacunarity curves for the sets in Figure 1.11 are illustrated in Figure 1.18.

The statistical behavior of $\Lambda(r)$ and the shape of the lacunarity curves can best be understood by recalling that:

$$Z(1) = \bar{s}(r) \qquad (1.15)$$
$$Z(2) = s_s^2(r) + \bar{s}^2(r) \qquad (1.16)$$

where $\bar{s}$ is the mean and $s_s^2(r)$ the variance of the number of sites per box. As a result,

$$\Lambda(r) = s_s^2(r)/\bar{s}^2(r) + 1. \qquad (1.17)$$

The lacunarity statistic is thus a dimensionless representation of the variance/mean ratio and is closely related to a number of statistics that have long been used in ecology (Ludwig and Reynolds, 1988).

From this relationship, and by examining the lacunarity curves in Figure 1.18, it can be shown that lacunarity is a function of:

1. The fraction, P, of sites that are occupied. As the mean number of occupied horizons $Z(1)$ goes to zero, Λ goes to ∞. Sparse sequences will thus have higher lacunarities than densely occupied sequences, for the same gliding box sizes.

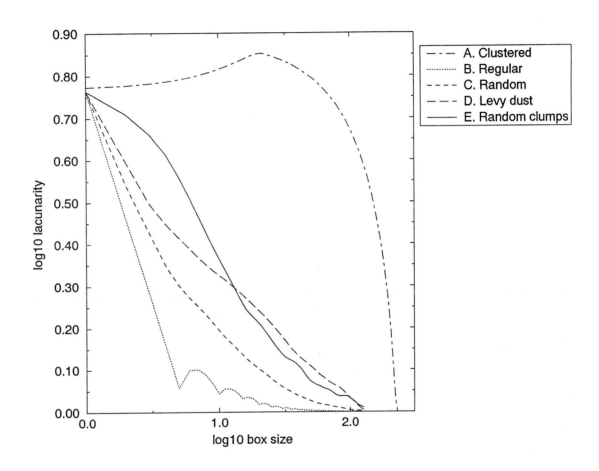

Figure 1.18: Lacunarity analysis of sequences in Figure 1.11.

2. The size r of the gliding box. In general, except for highly clustered sequences (e.g., sequence A in Figure 1.11) larger boxes will be more translationally invariant than smaller boxes; i.e., the second moment declines relative to the first. The same sequence will thus have lower lacunarities as the size of the boxes increase. For all sequences, since $Q(1,1) = P$, $Z(2)/(Z(1))^2 = P/P^2$, and $\Lambda(1) = 1/P$. This value is solely a function of the percentage of occupied sites and is independent of the overall size of the sequence and details of its geometry. A similar constraint occurs if the box is the size of the entire sequence; then the variance component of the second moment is 0 and $\Lambda(M)$ must equal 1. As a result, since all five sequences in Figure 1.11 have the same values of P and M, the Y- and X- intercepts of their lacunarity curves are identical.

3. The geometry of the sequence. For a given P and r higher lacunarity indicates greater clumping. Sequence A in Figure 1.11 is highly clustered, with a single large gap in the middle. For all $r << M$ most boxes are either mostly full or totally empty. As a result, the variance of box masses, and thus the lacunarity, is high over most of the range of box sizes. Notice that once the box size reaches that of the clumps, the curve declines very rapidly.

In contrast, the points in Sequence D are regularly distributed at a spacing of M/S. Once r is greater than this value, s would be constant at any location of the map, so the variance is zero. The lacunarity of a totally regular array is thus 1 for any gliding box size larger than the unit size of the repeating pattern.

Sequences B and C are intermediate cases. As expected, the lacunarity of the Lévy dust is much higher over all box sizes than that of the random sequence, since the Lévy dust is hierarchically clumped. The lacunarity curve of

the self-similar sequence is nearly linear. As described by Allain and Cloitre (1989), the lacunarity curve for self-similar fractals should be a straight line with a slope equal to $D - E$, where D and E are the fractal and Euclidean dimensions, respectively. The deviations from linearity in Figure 1.18 are due to the short length of the sequence. Analyses of larger sequences are much more linear (Allain and Cloitre, 1989; Plotnick, Gardner, and O'Neill, 1993). The random sequence, in contrast, forms a concave upward curve.

An examination of the lacunarity curve for sequence E, the randomly distributed clumps, demonstrates how lacunarity can be used to detect scales. The curve declines gradually to a break point at a log box size of about 0.6 (box size = 4), corresponding to the size of the clumps. It then declines more rapidly, with the concave upwards portion of the curve corresponding to the scales above that of random behavior.

Figure 1.19 is the lacunarity curve for the gamma-ray peaks in Figure 1.8. For comparison, the lacunarity curve for the same number of randomly distributed peaks is also shown. It can readily be seen that the gamma ray peaks are far more clustered, at all scales, than would be predicted from a random distribution and more closely approximate a fractal distribution.

Lacunarity analysis can also be applied to non-binary, quantitative data sets. Recall that lacunarity is also a measure of the variance/mean ratio of box mass. In this context, if the total mass (or measure, *sensu* Mandelbrot, 1983, 1989) is spread evenly over the entire set, then the variance, and thus the lacunarity, will be low. If the mass is concentrated at a few points, however, box mass variance and lacunarity will be high.

The application of lacunarity analysis to quantitative data can be illustrated by performing lacunarity analysis on the multifractal set produced by the binomial multiplicative process (Figure 1.13). Figure 1.19 illustrates the lacu-

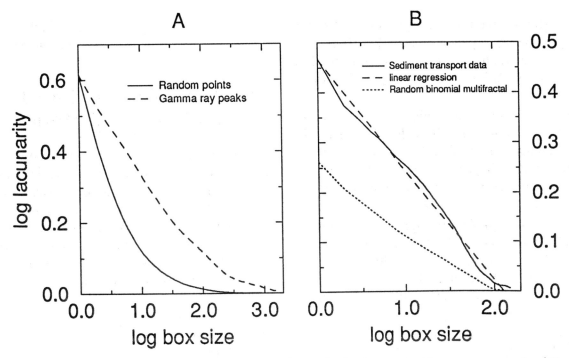

Figure 1.19: Lacunarity analysis of empirical and synthetic data. A. Gamma ray peaks (Figure 1.8) and random points. B. Sediment transport and binary multiplicative process (Figure 1.13 and 1.15).

narity curve for this sequence. As is the case for the fractals, the self-similar multifractal produces a linear lacunarity curve. Lacunarity can thus be used as a simple method for detecting multifractal structure in a data set.

Figure 1.19 also illustrates the results of a lacunarity analysis of the sediment transport data series of Figure 1.15. The plot is nearly linear, as a comparison with the corresponding least-square linear regression shows. This again supports the idea that this sequence can be mathematically modeled by the binomial multiplicative process.

Finally, as with binary data, lacunarity can be used to identify changes of scale within quantitative sequences. Plotnick et al. (in review) used multifractal binomial sequences to show that breaks in slope correspond exactly to changes in scaling behavior of the original distribution. This immediately indicates how multifractal models can be compared to empirical

data sets for the detection of scale dependent changes in spatial behavior.

1.8 Incommensurable Differences: Self-affine Fractals

One common aspect of objects viewed on a map, such as coastlines or rivers, is that the distance units or scales are the same in any dimension; e.g., kilometers north-south and kilometers east-west. In addition, our discussion so far of self-similar fractals implies that the rescaling has no preferred orientation; they are rescaled equally in all directions.

This is in sharp contrast to many types of data sets used by geologists. For example, a graph of a time series plots one variable (e.g. flow velocity) versus another (time) with completely different units. Alternatively, a cross-section might plot thickness (in meters), using some degree of vertical exaggeration, versus distance (in kilometers). The appearance of either graph is highly dependent on the plotting scales used. As a result, the concept of self-similarity, as discussed earlier, cannot be applied to these plots.

In general, rescaled graphs of theses types will "look the same" only if each axis is rescaled by different amounts, an *affine* transformation (Feder 1988). Sequences that look the same after an affine transformation are said to be *self-affine*, rather than self-similar.

The most familiar example to geologists of a self-affine sequence is classical Brownian motion (Voss, 1988; Feder 1989; Turcotte, 1992; Korvin, 1992; Plotnick and Prestegaard, 1993). Consider a sequence $B(t)$ (Figure 1.20A) whose initial value is zero ($B(0) = 0$). A random number $X(0)$, is now generated from a Gaussian distribution with a mean of zero and a variance of 1. This value is added to the initial value to get the second point in the sequence $B(1)$. This process is continued to produce the complete sequence $B(t)$. The sequence $X(t)$ of Gaussian random numbers $X(0,1)$ forms a white noise and the corresponding running sum of the white noise is a Brownian motion.

The Brownian motion in Figure 1.20A is 512 points long. Figure 1.20B shows points 100–151 after both y and x scales have been multiplied by a factor of 10. The appearance of this sequence is much more "hilly" than the original. In order to keep the sequence "looking the same," therefore, the y-axis must be rescaled by a lesser amount than the x-axis. In Figure 1.20C the x-axis has been multiplied by 10 but the y-axis by $10^{1/2}$. The graph now resembles that of Figure 1.20A; this is thus an affine transformation and Brownian motion is self-affine.

One difficulty with self-affine patterns immediately becomes apparent; how does one estimate the fractal dimension? As pointed out by Voss (1988), using the box-counting or Richardson methods requires arbitrarily fixing the x- and y-coordinates. As we will see later (next

section; Chapter 3) analyzing self-affine patterns requires a different set of techniques.

1.9   The Colors of Time: Fractal Noises and Motions

White noise and Brownian motion, as discussed above, are two familiar approaches to modeling a sequence of values in time. White noise represents a series where successive values are independent but are drawn from a common frequency distribution; i.e., the series is stationary and there are no short- or long-term correlations. Brownian motion, on the other hand, is a non- stationary Markov process where successive values are correlated, but the directions and magnitudes of change between successive values are not. White noises and Brownian motions are closely related, with the increments of a Brownian motion being a white noise.

A customary technique for the analysis of time series is Fourier analysis (Davis, 1986). Fourier analysis represents a series as the sum of a series of sinusoidal curves of different frequencies, amplitudes, and phase angles. The results of a Fourier analysis are often portrayed in the form of a power spectrum, where the power (= amplitude squared; = sum of squared Fourier components) of a sinusoidal component is plotted against its frequency (f). The larger the power at a particular frequency, the larger its contribution to the total variance in a series.

When a white noise is analyzed using Fourier techniques, the resulting power spectrum is more-or-less flat; i.e., the power is independent of frequency (Figure 1.21) . Since all frequencies contribute equally to the variance in the series, the analogy is made to white light, and thus the term "white noise "(Schroeder, 1991).

In contrast, spectral analysis of a Brownian motion shows a rapid decrease with frequency, proportional to f^{-2} (Figure 1.22; since this is a single realization of a random process, β does not equal exactly 2). Power thus declines in a *Continued on p. 24*

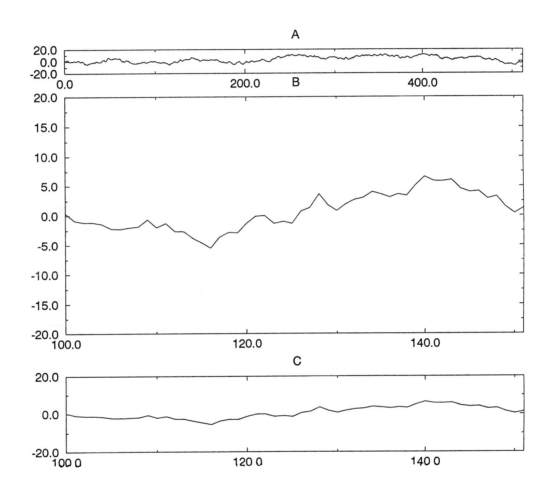

Figure 1.20: Self-affinity of Brownian motion.

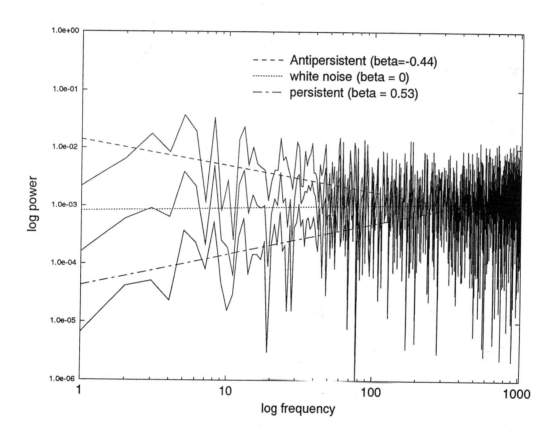

Figure 1.21: Fourier spectra for white and fractional Gaussian noises.

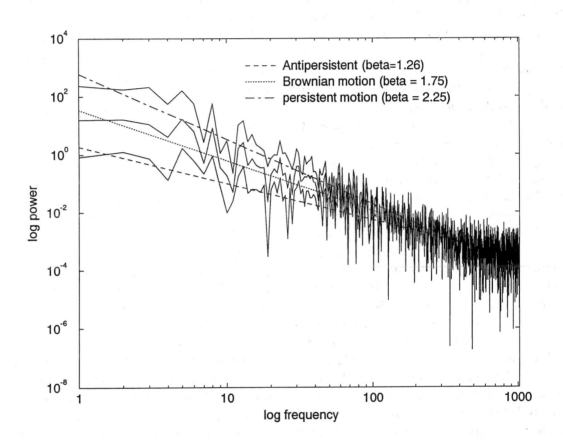

Figure 1.22: Fourier spectra for Brownian and fractional Brownian motions.

self-similar manner with frequency, with higher frequencies contributing proportionally less to the total variance. Schroeder (1991) terms patterns that show spectra similar to that of Brownian motions "brown noises."

Analysis of many empirical patterns have revealed that a decline of the power spectrum proportional to some power frequency is ubiquitous (Mandelbrot and Wallis, 1969, Voss, 1988; Feder, 1988; Schroeder, 1991). These sequences have been therefore been termed $1/f^\beta$ noises or, more simply, $1/f$ noises. For white noises, $\beta = 0$, whereas for brown noises $\beta = 2$. Note that the β value for the Brownian motion equals that of the white noise (+2).

Empirical values of β, however, have rarely, if ever, been found, to equal 0 or 2. For example, Figure 1.23 represents a spectral analysis of the gamma ray spectra described above. The power declines proportial to $f^{-0.57}$. Other studies of electric well-logs have shown similar results (Hewett, 1986; Turcotte, 1994a; Tubman and Crane, 1995). Notice that this spectrum has more power in the low frequency ("red") end of the spectrum than does a white noise; as a result, spectra with values of $0 \leq \beta \leq 1$ are known as *pink noises* (Schroeder, 1991). Similarly, Schroeder refers to power spectra with $\beta > 2$ as *black noises*. Notice that $1/f$ noises are a power law phenomenon and are thus fractal; i.e., the amplitude of the wave forms scale in a self-similar way with frequency. Specifically, they they scale with a factor r with frequency and a factor $r^{-\beta/2}$ for amplitudes (Schroeder, 1991) and are thus self-affine.

A mathematical model that encompasses white, pink, brown, "brownish" $(1 > \beta < 2)$, and black noises was proposed in the 1960's by Mandelbrot and his co-workers (Mandelbrot and Van Ness, 1968; Mandelbrot and Wallis 1968, 1969a). Recall that a Brownian motion is the running sum of a white noise or, conversely, that the intervals between steps of a Brownian motion are a white noise. In addition, recall that the successive values of a white

noise are independent; the probabilities of an "up" or "down" value are equal.

In many natural situations, however, we would not expect to find incremental values of a processs to be uncorrelated; instead, we find that positive increments tend to follow positve increments and negative increments will follow negative increments. For example, days with high and low river flows are not independent, days with high flows tend to follow days with high flows and vice versa. These types of sequences thus show *persistence*. A white noise would thus not be an appropriate model for this situation.

Instead, Mandelbrot, Wallis, and van Ness introduced the concept of a *discrete fractional Gaussian noise* (dfGn). A fractional noise resembles a white noise by being drawn from a Gaussian distribution; i.e., over time both white and Gaussian noises would have the same frequency distribution. The differ, however, in that the successive values of the dfGn are *not* independent; instead, successive values are either positively or negatively correlated. If they are positively correlated, they show persistence; if they are negatively correlated (negative follow positive and *vice versa*) they show *antipersistence*. Examples of white, persistent, and antipersistent noises are shown in Figure 1.24 (all three were generated using the same random seed; see Saupe, 1988 for a discussion of the algorithms used to generate the noises). Notice that the persistent noises show long periods where their values are either above or below the mean of zero, while antipersistent noises show rapid fluctuations above and below the mean.

Discrete fractional Gaussian noises possess $1/f^\beta$ power spectra. For antipersistent noises, $\beta < 0$, whereas for persistent noises $0 < \beta < 1$ (Figure 1.21). As a result, persistent noises are a mathematical model for "pink noises," such as the gamma ray log in Figure 1.23. This has led to their use as a technique for the simulation of logs, as well as for interpolation between logs (Hewett, 1988; Turcotte, 1994a; Tubman and

Continued on p. 28

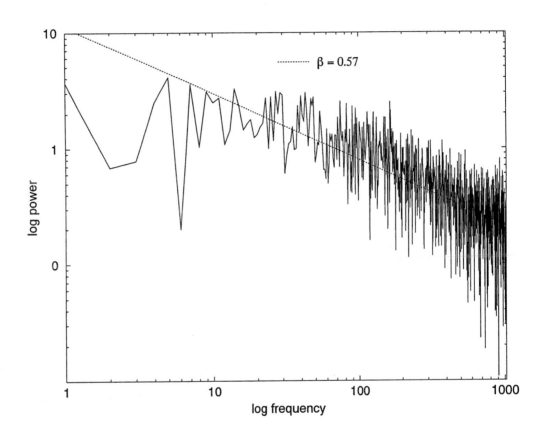

Figure 1.23: Fourier spectrum for gamma-ray log.

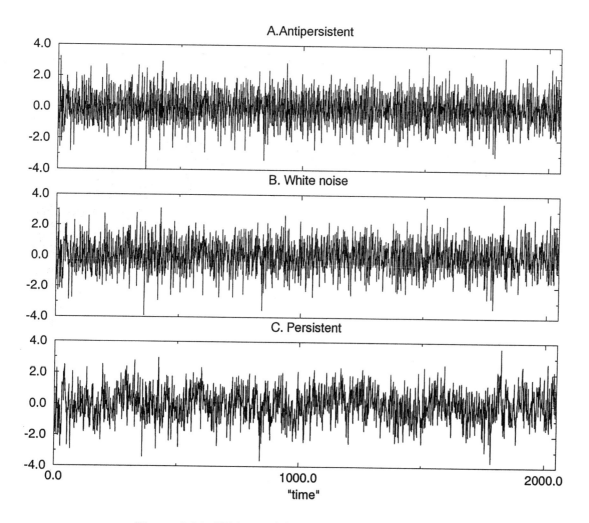

Figure 1.24: White and fractional Gaussian noises.

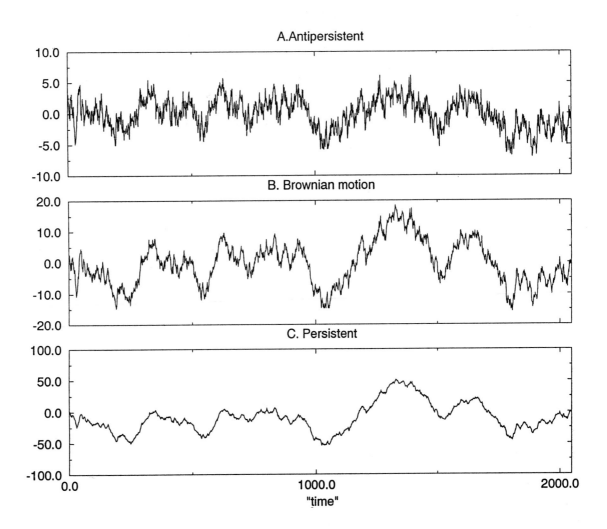

Figure 1.25: Brownian and fractional Brownian motions.

Crane, 1985).

Just as white noises can be "integrated" to produce Brownian motions, discrete fractional Gaussian noises can be summed to produce *fractional Brownian motions* (fBm). Figure 1.25 shows the motions resulting from summing the noises shown in Figure 1.24. Because the same random number seed was used to produce all three curves, their overall shapes are very similar. However, two remarkable differences exist:

1. The range of the motions increases tenfold from the persistent to the antipersistent motion.

2. The antipersistent motion is much "rougher," in the sense of a more jagged appearance.

These differences are due to the persistence or antipersistence of the increments. In other words, persistent motions have much longer trends and reversals of direction than antipersistent motions.

The Fourier spectra of the motions are shown in Figure 1.22. Notice that the β values are all greater than 1. Again, these sequences can be used as a mathematical model for brown and black noises. A program demonstrating these types of motions can be found on the diskette included with these Notes.

Finally, note the similarity of these patterns to topographic cross-sections. One of the earliest use of fractals was to generate synthetic topographies using 3-D versions of fractional Brownian motions (Mandelbrot, 1983; Voss, 1988; Klinkenberg and Goodchild, 1992; Turcotte, 1992; Lavallée et al., 1993). One of the most successful uses has been to model topographic variability of the sea-floor (Malinverno, 1989, 1995; Mareschal, 1989).

Chapter 2

INTRODUCTION TO NONLINEAR MODELS

Gerard V. Middleton
Department of Geology
McMaster University
Hamilton ON L8S4M1, Canada

with a contribution by
David M. Rubin
U.S.Geological Survey
345 Middlefield Road, MS 999
Menlo Park CA 94025 U.S.A.

2.1 DETERMINISTIC vs STOCHASTIC MODELS

In the following discussion we will use the term "model" as a short form of "mathematical model", that is, a set of equations that are designed to represent some aspect of the real world. We recognize that other types of models are possible, e.g., conceptual models that cannot be exactly quantified, or physical models, that are generally models built of real materials, but at a smaller scale than the real phenomenon. We can recognize two different extreme types of mathematical models:

- **Stochastic models**. In such models it is supposed that the phenomena show some definite regularities (whose origin need not be well understood), more or less modified or masked by an irregular, unpredictable

variation. The simplest possible example might be a linear regression model:

$$y = a + bx + \epsilon \qquad (2.1)$$

y might be mean grain size (in Phi units), and x might be distance downstream in a river. a and b are parameters, whose value might be determined for a particular river by a least squares technique. ϵ is "error", or natural variation caused by deterministic or stochastic processes not addressed by the other terms in the equation. We assume that the predicted value for grain size, given by

$$\hat{y} = a + bx$$

will differ from the value actually measured by a quantity ϵ which varies in a way that is unpredictable (except statistically). We might assume that the average value of ϵ is zero, and that it has a certain statistical distribution (probability density function)—often we assume a Normal distribution—characterized by certain parameters, for example, the standard deviation of the "error".

The "error" term is frequently called a *random variable*, and the introduction of such a term (or terms) into a model changes the model from a deterministic to a *stochastic model*. Note that there is really a gradation from models which are purely deterministic (i.e., "error" is a very small part of the total variation), through models which are partly determinist but have large error terms, to models based entirely on stochastic variables (such as random walk models).

- **Deterministic models**. The archetypical deterministic model is provided by the application of Newtonian dynamics, for example, to the motion of a projectile. Such a model is generally called a *dynamical model* and consists of one or more differential equations, for example

$$dz/dt = -gt + w_0 \qquad (2.2)$$

This equation describes the vertical position z of a projectile, thrown upwards in a gravitational field g at an initial velocity w_0, as a function of time t. In this case, there is a well established theory that predicts the form of the equation, and (under certain, well understood conditions) the predictions are expected to correspond exactly with reality. Thus no error term is included in the equation. If actual experiments were performed there would, indeed, be some errors in evaluating g, t, and w, and errors in dz/dt would be expected to be linear combinations of these errors in the other variables.

In present usage, the term "dynamical model" is not restricted to applications of the laws of Newtonian mechanics. It is applied to any model that describes the way in which a system changes with time. In its purest form, a dynamical model is also deterministic, but it may be made stochastic by including a random variable. The model may deal with continuous systems, in which case it generally consists of a set of differential equations, or it may describe systems that change in discrete steps from one state to another, in which case it consists of a system of difference equations, or

"mappings" of the state of the system at one time into its state after the next interval of time.

To some extent, the distinction between continuous and discrete systems is arbitrary. For example, one can write two versions of the *logistic equation*, which describes the growth of a population as it approaches some naturally limiting value, scaled to $x_{max} = 1$:

- continuous:

$$dx/dt = rx(1 - x) \qquad (2.3)$$

- discrete:

$$x_{n+1} = rx_n(1 - x_n) \qquad (2.4)$$

The second equation might be regarded as a finite difference approximation of the first—or alternatively, as a proper description of a phenomenon that is inherently discontinuous (e.g., the annual change in a population which produces offspring only once a year). We will see, however, that the change from a continuous to a discrete form may produce a marked difference in the behavior of the system.

The traditional way to distinguish between the two types of systems was that deterministic systems displayed smooth, regular behavior, whereas stochastic systems produced erratic, irregular behavior. In 1963, however, Lorenz showed that a simple deterministic model of convection in the atmosphere produced highly erratic behavior, that was predicatble only for short times into the future. In 1976 May pointed out that the simple difference form of the logistic equation displayed amazingly complex behavior. These, and some other examples, have convinced scientists that it is not easy to distinguish systems that are truly stochastic from those that are deterministic, yet apparently unpredictable. As Ford (1983) pointed out, the archetypical stochastic process is tossing a coin—yet, in principle the result (head or tail) is produced by a dynamical system which in theory (and if we neglect some small effects due to air resistance, etc.) is completely deterministic.

2.2 NONLINEARITY

A dynamic model attempts to apply a set of governing equations, such as Newton's laws, coupled with constitutive equations, such as the flow law for a viscous fluid, to a problem of scientific interest. In order to do this, it is necessary to specify not only the governing equations, but also the boundary and initial conditions, and then attempt to integrate the equations. This is generally a difficult problem, even if the governing equations are linear, because the boundary and initial conditions can be very complicated. If the governing equations are linear, as is the case for example in the heat-flow equation or the wave equation, then it is commonly (though not always) possible to find many different solutions, which can be added together to satisfy the boundary and initial conditions. This is the basis for making use of Fourier series to solve such equations. Many of the governing equations are, however, nonlinear—and these techniques cannot be applied. Rather than abandon them, much of classical physics and geophysics has sought to extend their application by reducing the nonlinear equations to a simpler, linear approximation. Only since the widespread availability of computers has it become common for scientists to attempt to study nonlinear dynamical systems by obtaining numerical solutions.

Nonlinear dynamics is simply the study of systems whose governing equations are nonlinear. It seems, therefore, that it should differ from linear dynamics only quantitatively: the equations will be more complicated and harder to solve. What has made nonlinear dynamics a new science, and not simply a more difficult branch of the old dynamics, has been the realization that nonlinearity causes behaviour that is not only extraordinarily complex, even for very simple sets of dynamic equations, but also qualitatively different from any behaviour shown by linear systems.

The dictionary definition of **chaos** is "total disorder or confusion" (American Heritage). The existence of natural chaotic states has long been recognized, or at least postulated, by sciences that invoke random variables or stochastic models. Statistical mechanics and statistical theories of turbulence are just two well-developed examples. The term "chaos," however, was rarely used to describe these phenomena. Its popularity as a scientific term is new and probably stems from the title of a paper by Li and Yorke (1975).

Before the development of the new ideas about chaos, highly irregular behavour was ascribed to the action of one or more random variables, or to the influence of a very large number of (unknown) variables—which is essentially the same thing. Random variables are either simply postulated to be completely random, or their average properties are thought to result from summing the action of a very many separate variables, which are individually unknown and perhaps even in principle unknowable. Random behaviour of this type is essentially the product, either of inherent ignorance, or of very large numbers of degrees of freedom in the system.

In contrast, it is now known that some types of highly irregular behaviour (deterministic chaos) can result from the action of a small number of variables in a system of nonlinear governing equations. In principle, therefore, it is completely deterministic—and the same pattern of chaos could be reproduced by the same system if the initial conditions were known with perfect accuracy. Of course, this is a very big "if"—in reality, most nonlinear systems, even those consisting of only a few sets of equations, cannot be solved analytically. Numerical solutions must be obtained using computers whose precision is necessarily limited. So even if the initial conditions were known perfectly, the intermediate steps in the calculation could only be carried out by numerical approximation, and thus states of the system remote from the starting state could be known only approximately.

But the real situation is worse. The main distinguishing characteristic of chaotic systems is that "errors" cumulate exponentially. Such systems display "sensitive dependence on initial conditions" (Ruelle, 1979). Linear systems, even when they incorporate stochastic variables, are not like that. We learn in elementary physics laboratories to estimate probable errors for our calculated results by considering the effects of the small errors that we are likely to make in each of our measurements. With reasonable luck, some errors "cancel out"—but even in the worst case we assume they may simply add up. In nonlinear chaotic systems, this is not the case: very small "errors" (differences in initial conditions) soon give rise to very large differences in computed results—and this results from the dynamics of the system not from computational error.

Sensitive dependence on initial conditions was known to Poincaré and other mathematicians by the turn of the century, but its impact on science was not felt until the work of Lorenz (1963) and Hénon and Heiles (1964). In both cases it was rediscovered "experimentally" in the process of investigating numerical solutions of sets of equations thought to be highly simplified models of natural phenomena (in meteorology and astronomy respectively).

Sensitive dependence on initial conditions does not necessarily imply that future states of chaotic systems are totally unknown. In dissipative systems (and that includes most systems of interest in the earth sciences) future states converge on (though in theory they never quite reach) a **strange attractor**—a set of solutions, which though it occupies only a part of the possible solution space, is called "strange" because it is not a simple point, line or surface in that space, but instead is a set of points with fractal properties. A classic example was described by Lorenz (1963), though the term itself was first used by Ruelle and Takens (1971), and not clearly defined. The result is that some authors use "strange" to refer to the chaotic properties of such attractors, and some use the term to refer to their fractal properties. For examples of attractors that are fractal, but not chaotic, see Ott (1993, p.205).

Sensitive dependence can be measured by observing the rate at which trajectories starting from points, originally located close to each other on (or near) the attractor, diverge with time. In the early stages, the divergence is exponential, and can be quantified by a series of **Lyapunov exponents** (there is more than one, because divergence can be measured in different directions). If at least one Lyapunov exponent is positive, then the system can be regarded as chaotic. Of course, exponential divergence cannot continue for long, because trajectories are confined to the attractor, which is only a small part of the total state space, but trajectories beginning from nearby positions, may end up almost anywhere on the attractor. The distance between two trajectories at time t is given by

$$\Delta x_t = \Delta x_0 e^{\lambda_1 t}$$

where λ_1 is the largest Lyapunov exponent. When Δx approaches the size of the attractor, no useful prediction can be made (other than a prediction based on the statistical properties of the attractor itself).

2.3 NONLINEAR MODELS: EXAMPLES

A few nonlinear models have become "classic" and are referred to frequently in the literature. Unfortunately, there is no single reference work that describes them all — some of the best sources of information are Holden (1986), Moon (1992) and Peitgen et al. (1992). In what follows we give brief outlines and some further references for the most frequently cited models. For others, see Appendix 1.

2.3.1 Oscillators

Earth scientists generally learn about simple oscillating systems, such as the pendulum or vibrating spring, in freshman physics, but they often do not realize the extent to which these systems have become a paradigm for the explanation of any system that varies through time, in a "uniformitarian" (non-evolutionary) manner. Astronomical systems, and the effects they produce, such as tides and climatic variations, can be considered to be examples of large-scale oscillating systems. The stratigraphic record of eustatic and tectonic cycles can also be interpreted in this way—though the reality and nature of stratigraphic cycles is a subject of long-standing geological controversy. If we are to have any hope of understanding such complex systems, we must begin by considering simpler examples.

As an example of a simple linear dynamic system consider the oscillator

$$\frac{d^2x}{dt^2} = -\omega^2 x \qquad (2.5)$$

This represents the equation of motion of a mass sliding to and fro at the end of a spring, over a horizontal frictionless surface (or alternatively, and perhaps more realistically, a mass oscillating on the end of a spring in a gravity-free space module). The spring exerts a force proportional to the displacement x, and the coefficient of proportionality, or *spring constant* is ω^2. Two integrations would show that x is nonlinearly related to t, but this equation is nevertheless classified as a linear differential equation because each term contains only x (itself a function of t) or its time derivatives to the first power. Note also that it is possible to write this second-order equation in a different form, as a system of two (linear) first order equations:

$$\begin{aligned} \frac{dx}{dt} &= u \\ \frac{du}{dt} &= -\omega^2 x \end{aligned} \qquad (2.6)$$

It is conventional to display a dynamical system in this way.

The system can easily be made nonlinear by adding another force to the equation. For example, the **van der Pol equation** is

$$\frac{d^2x}{dt^2} = (\epsilon - x^2)\frac{dx}{dt} - \omega^2 x \qquad (2.7)$$

which can be written as the two first order equations

$$\begin{aligned} \frac{dx}{dt} &= u \\ \frac{du}{dt} &= (\epsilon - x^2)u - \omega^2 \end{aligned} \qquad (2.8)$$

We see that these equations are nonlinear because they contain terms like $x^2(dx/dt)$—powers of x or products of x with its derivative (in this case, both).

The simple oscillator shows simple harmonic motion; the van der Pol oscillator shows somewhat more complicated but still regular, predictable motion. The best way to visualize the motion is not to show what are generally called "trajectories", that is, the way that x varies with time, but instead to plot x against its derivative dx/dt, i.e., against velocity. For a constant mass, this is also equivalent to plotting the position against the momentum, or showing the trajectory in **phase space**. Time is not shown directly, but with time the system traces out a single continuous curve in phase space. More generally, this type of diagram shows indirectly how one state of the system follows another in time, so the space may be referred to as **state space**.

The behavior of oscillators can be made yet more complicated by introducing a periodic driving force. As an example, consider the

Duffing equation, which describes an electrical circuit with a nonlinear inductor:

$$\frac{d^2x}{dt^2} = -k\frac{dx}{dt} - x^3 \qquad (2.9)$$

We can introduce a periodic driving force by adding the term $b\cos t$ on the right hand side of the equation, and by a few simple algebraic manipulations arrive at the following nonlinear system:

$$\frac{dx}{dt} = y \qquad (2.10)$$

$$\frac{dy}{dt} = -ky - x^3 + b\cos 2\pi z \qquad (2.11)$$

$$\frac{dz}{dt} = 1 \quad (\mathrm{mod}\,1) \qquad (2.12)$$

The last of the three equations is just a way of expressing that the driving force is periodic with time. A plot of x again y at regular time intervals readily yields a Poincaré map of such a system.

Driving a dynamical system by applying a regular force may modify the behavior of the system in ways that are not easy to predict: it may produce only small subtle effects, or very large oscillations (this is the phenomenon of *resonance* due to a near approach of the driving frequency to one of the natural frequances of the dynamic system), or highly irregular, chaotic behavior. The regular, predictable behavior has been studied for many years: Duffing published (in German) in 1918, and van der Pol in 1927, and an excellent summary of this earlier work is given by Stoker (1950). But investigation of the chaotic behavior of such systems had to wait on the development of digital computers, and even then was resisted by some authorities (e.g., see Moon, 1992, p.148–150).

The chaotic behavior of the forced Duffing equation has been well documented by Ueda (1980; see also Thompson and Stewart, 1986, Chapters 5 and 6; Moon, 1992, Chapters 4 and 7), and similar studies have been made for the forced pendulum (e.g., Baker and Gollub,

1990). Moon studied the fractal dimension of the Poincaré maps of the strange attractor for a forced Duffing system with a range of damping coefficients $k = 0.06 - 0.19$, and a forcing frequency of 8.5 Hz. He showed that the fractal dimension of the attractor section did not depend on the phase, i.e., on the particular Poincaré section studied, but it did depend on the damping: it was only slightly larger than 1.0 for highly damped systems, but as high as 1.6 for weakly damped systems.

For a more complete (yet elementary) discussion of nonlinear oscillators see Thompson and Stewart (1986), Baker and Gollub (1990), and Strogatz (1994). See also the discussion in Chapter 5, and the entries in Appendix I under **Attractor, Duffing equation, Oscillator, Van der Pol equation.** Moon (1992) gives many examples of mechanical oscillating systems that show chaotic behavior.

2.3.2 The Lorenz System

The classic chaotic system is that originally described by Lorenz (1963). Lorenz derived his equations from a Fourier series expansion of the full equations for thermal convection (for accounts see Bergé et al., 1984, Appendix D; Jackson, 1990, v.2, Chapter 7; Middleton and Wilcock, 1994, p.404–407; Peitgen et al., 1992, Chapter 12; Strogatz, 1994, Chapter 9; Thompson and Stewart, 1986, Chapter 11; Turcotte, 1992, Chapter 12; and Tritton, 1988, Chapters 17 and 24). He retained only one equation for the vertical component of velocity, and two equations for the distribution of temperature. The resulting set of nonlinear differential equations is

$$\frac{dX}{dt} = -Pr(X - Y)$$

$$\frac{dY}{dt} = rX - Y - XZ \qquad (2.13)$$

$$\frac{dZ}{dt} = XY - bZ$$

In these equations, Pr is the Prandtl number, which is the ratio of kinematic viscosity over the thermal diffusivity: it expresses how fast momentum is diffused compared with heat. r is the ratio of the Rayleigh number over the critical value required to initiate convection. The Rayleigh number is the ratio of two times: the time that a volume of fluid takes to cool, and the time that it takes to rise a characteristic distance—if the volume loses heat faster than it is rising due to buoyancy, convection will cease. X, Y, and Z are not spatial coordinates. X is proportional to the amplitude of the linear component of vertical velocity, Y is proportional to the amplitude of the linear component of temperature, and Z is proportional to the amplitude of the single harmonic of temperature retained in the analysis. $b = 8/3$ and is related to the aspect ratio $(1/\sqrt{2})$ of the "box" within which convection takes place.

The equations have no general analytical solution but can easily be integrated numerically. The behavior is generally shown by plotting the trajectories in state (X, Y, Z) space, using $Pr = 10$ and $r = 28$ (Figure 2.1). It can also be shown by plotting a single variable against time (Figure 2.2), or by showing the actual trajectory in "real" space, as opposed to state space (Figure 2.3).

What is actually happening as a trajectory moves through state space is nicely illustrated by Figure 2.4, taken from Kellogg and Turcotte (1990). The small diagrams inset on the state-space projection (on the YZ-plane) show the instantaneous pattern of streamlines typical of each section of state space. These patterns are constantly changing with time, from one cell to two cell convection, and from clockwise to anticlockwise movement in a cell. This probably illustrates better than any simple state-space plot, just how complicated the motion is in this particular regime. Figure 2.3 also illustrates nicely how such chaotic motion produces very effective mixing throughout the region of convection (for further discussion of mixing by

convection, see Kellogg, 1992).

The Lorenz attractor has been shown to have a fractal (correlation) dimension of 2.06—that is, it is very nearly, but not quite two-dimensional. This is because the trajectory in state space lies on many different "sheets" of the Lorenz attractor (in fact, there are an infinite number of such sheets). Each trajectory in state space is, of course, smooth but the intersection of the trajectories with a plane (a Poincaré section) does not define a smooth curve and has fractal properties.

The Lyapunov exponents have been estimated by Wolf et al. (1985) and Briggs (1990), for $Pr = 16$, $r = 45.92$ (not the standard values quoted earlier) to be $\lambda_1 = (1.5 - 1.6)$, $\lambda_2 = 0$, $\lambda_3 = -(21 - 22.5)$. The positive exponent indicates that nearby trajectories diverge exponentially on the surface of the attractor.

The largest Lyapunov exponent determines how fast trajectories diverge, and therefore how far into the future it is possible to make predictions about the behavior of the system. When the distance separating originally close trajectories is as large as the attractor itself, all predictive power is lost (even though the system is in principle completely determinate, rather than stochastic). The large negative third Lyapunov exponent (and the negative sum of the three exponents) indicate that the system as a whole is strongly dissipative. Trajectories starting away from the attractor are drawn rapidly towards its surface, and a small volume in state space contracts in size as it moves along, even as it is stretched (by exponential divergence in one direction), because it is contracting more rapidly in the third, orthogonal direction. Exponential divergence does not continue indefinitely in any direction because the attractor as a whole is confined in a part of state space.

The extent to which trajectories diverge generally varies locally on the attractor surface, and therefore so does the range of possible prediction. The future may be much more predictable from some starting positions than from others—

Figure 2.1: The Lorenz attractor, for $r = 28$ from Moon (1992, p.41).

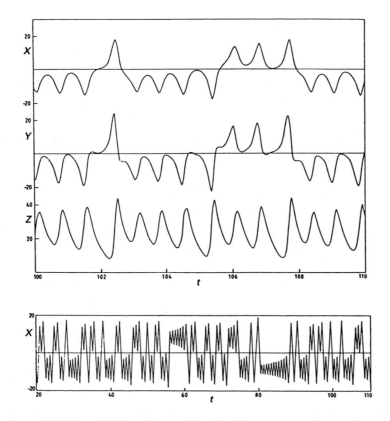

Figure 2.2: A plot of X, Y and Z against t for the Lorenz equations, for $r = 28$, Tritton (1988, p.397). The lower box shows a more extended time series for X.

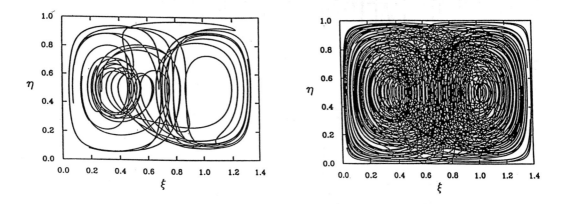

Figure 2.3: A pathline for Lorenz convection with $r = 28$, from Kellogg and Turcotte (1990), for times 10 (left) and 100 (right). η and ξ are vertical and horizontal directions.

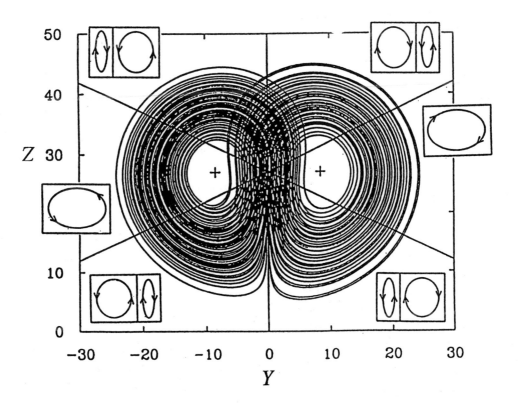

Figure 2.4: Patterns of convection associated with regions of state space for the Lorenz attraction, with $r = 28$, from Kellogg and Turcotte (1990).

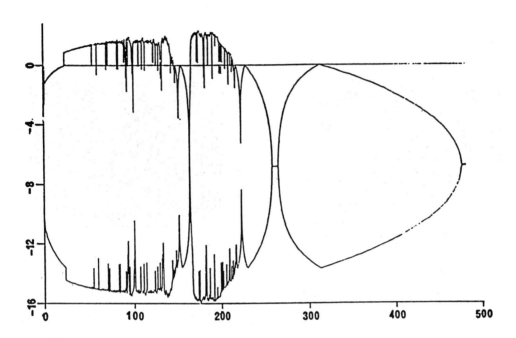

Figure 2.5: Lyapunov exponents for the Lorenz system in the range $0.1 < r < 520$, determined at r increments of 0.1. From Frøyland and Alfsen (1984).

and this has been discussed with particular reference to the Lorenz attractor and weather forecasting by Palmer (1993).

We have described the Lorenz system in its chaotic regime—but for different values of the parameter r, the system may show much different behavior. For $r < 1$ there is a single stable attracting point at $X = Y = Z = 0$ (i.e., no convection), and for $1 < r < 24.74$ there are two attracting points. The chaotic regime is found in the range $24.74 < r < 30.1$, and above this value there are "windows of periodicity" (ranges of r where the system oscillates in a regular, periodic manner) interspersed with chaos. Figure 2.5 is a diagram showing how the three Lyapunov exponents vary with r for the Lorenz equations with the standard parameters. The positions of the stable periodic windows are indicated by the ranges of r where the largest (non-zero) Lyapunov exponent becomes negative. Note that, from the way Lyapunov exponents are defined, one of the exponents in a

system of differential equations is always zero.

Though the Lorenz equations were derived for a simplified model of atmospheric convection, they have been found to describe several other physical systems, such as a water wheel (Kolář and Gumbs, 1992; Strogatz, 1994) and convection in a circular pipe (Moon, 1992; Tritton, 1992; Jackson, 1990, p.151–155). Similar systems have been described in lasers (Haken, 1985, Chapter 8) and in a model of the earth's magnetic dynamo (Ito, 1980; Jackson, 1990, p.155–162).

2.3.3 The Rössler System

This chaotic system is described in many texts and papers because it is based on a very simple set of nonlinear equations.

$$\frac{dX}{dt} = -Y - Z$$
$$\frac{dY}{dt} = X + aY \qquad (2.14)$$

$$\frac{dZ}{dt} = b + Z(X - c)$$

There is only a single nonlinear term (in the third equation), but this is sufficient to produce chaos in a continuous system with at least three variables. The usual choice of parameters is $a = 0.398$, $b = 2$, and $c = 4$, but many others are also used. This system was invented by its author simply to show chaos: it was not derived from any physical model. For $a = 0.15$, $b = 0.2$, and $c = 10$, Wolf et al. (1985) determined that the Lyapunov exponents are $\lambda_1 = 0.13$, $\lambda_2 = 0$, $\lambda_3 = -14.1$. The attractor dimension is 2.01, even closer to two-dimensional than the Lorenz attractor. In fact, the Rössler attractor is often likened to a single folded band of trajectories (see Figure 4.1; for practical instructions on how to make a paper model, see Peitgen et al., 1992).

2.3.4 Other differential systems

A large number of other systems have been described by other writers. The diskette accompanying these notes includes demonstrations of the Lorenz, Rössler, and several other attractors. Several sources of inexpensive software which demonstrate numerous other attractors are listed in Appendix II.

In one study of particular interest, Sprott (1993, 1994) programmed a computer to search systematically through very general sets of differential and difference equations, looking for those that displayed chaotic behavior, as revealed by a positive largest Lyapunov exponent. He gives the equations and Lyapunov exponents of 19 previously undescribed, simple, three-dimensional systems that are chaotic for some range of the parameters. It is interesting, but perhaps fortuitous, that most of the attractors have a fractal dimension only a little larger than 2. An example of an attractor with a relatively large dimension (about 2.17) is his "Attractor B", and a program to generate this attractor is given in the diskette accompanying

these notes.

Note that in a differential system, one of the Lyapunov exponents is always equal to zero. There is also a theoretical relationship between the exponents and the dimension of the attractor: the **Kaplan-Yorke dimension** is defined by

$$d_L = k + \frac{\sum_i^k (\lambda_i)}{-\lambda_{(k+1)}} \qquad (2.15)$$

where the exponents are ordered from largest to smallest, and k is defined by the requirement that

$$\sum_i^k \lambda_i > 0$$

For the Rössler example, $k = 2$ and $d_L = 2 + 0.13/14.1 \approx 2.01$ The Kaplan-Yorke dimension is thought to be a lower bound on the capacity (box-counting) dimension.

Three equation systems are obviously the best for visual demonstrations of chaotic trajectories, because they can still be plotted (and even examined in stereographic projection). But it should be remembered that the classic Lorenz and Rössler attractors are somewhat special. The fractal dimension is so close to 2 that the attractors seem to be two dimensional surfaces, with the trajectories moving over them. Poincaré sections that cut through these surfaces seem to be curves; but in fact they have a complex, fractal structure that is too fine to be seen in the usual demonstrations. For a system that displays the fractal stucture of the attractor more vividly, see the illustrations of the forced Duffing oscillator (Ueda attractor) given by Ueda (1980), Thompson and Stewart (1986, p.102–107), or Sprott (1993, p.298–304).

Of course, higher dimensional systems also exist. Rössler provided an example, which he called his "hyperchaos" system (see Holden, 1986, p.26).

2.3.5 The Logistic Map

We have already presented the continuous (Equation 2.3) and discrete (Equation 2.4) form of the Logistic equations. The continuous form can be integrated exactly, and produces a single value of x for any chosen value of the parameter r. The same result can be produced if the equation is integrated by numerical methods, using a sufficiently small finite time difference. But if a larger time difference is used the equation "blows up," producing numerical chaos. Such behavior was well known to numerical analysts long before nonlinear dynamics became popular.

What was not understood before the work of May (1976) and others, is just how complex and interesting the behavior of the discrete form of the logistic equation really is. There are now known to be many such difference equations (generally called *maps*, as opposed to the *flows* produced by continuuous systems) which display chaotic behavior (see Appendix I for other examples). The logistic system has been studied, not as an approximation to the continuous form, but as a very simple map that displays many of the general characteristics of chaotic systems. For example, the text by Devaney (1989) is devoted almost exclusively to an examination of this system.

We restrict our discussion to a presentation of the famous diagram (Figure 2.6) showing the way the possible solutions for x change, as r varies from 3.5 to 4. This diagram is constructed by repeatedly solving the equation on the computer—this is done first about 100 times, to allow the solution to converge on any stable solutions (i.e., so called "fixed" or attracting points). These 100 "transients" are discarded before the first points are plotted on the diagram. It turns out that for $0 < r < 3$, there is always a single fixed point, for $3 < r < 3.449\ldots$ there are two fixed points, then (in rapid sucession as r increases) there are $4, 8, 16, \ldots$ until, at about $r = 3.5699\ldots$ there are no regular, cyclic solutions, and x varies apparently randomly between certain limits (as shown by the cloud of dots in the figure). The doubling in the number of fixed points is an example of *bifurcation*—a

type of behavior that has been much studied by mathematicians, and which often (but not always) preceeds the onset of chaos. It is known as the "period-doubling route to chaos," and is displayed by physical systems, such as thermal convection in fluids (Bergé et al., 1984, p.210–213; Strogatz, 1994, p.374–376), and by dynamical systems such as the Rössler system, as well as by one-dimensional maps.

The stability or instability of the logistic system, may be quantified, as for continuuous systems, by calculating the Lyapunov exponent: there is only one, since the map is one-dimensional (Figure 2.7). As for the Lorenz system, it can be seen that after chaos sets in, there are still windows of periodic behavior, where there are also further examples of bifurcation sequences very similar to the one for $r < 3.57$. The fact that similar patterns are repeated at difference scales within the map is one aspect of the fractal structure of such maps. Another is that it has been shown that, for $r = 4$, the distribution of points (occupying the entire range $0 < x < 1$) is strictly fractal, with the geometry of a random Cantor dust.

We may wonder why it is that differential systems can show chaos only if the dimension is three or larger, but even one-dimensional maps can be chaotic. Part of the answer is that any two-dimensional map can be thought of as a section through a three-dimensional flow—so it is not surprising that such systems can show chaos. One-dimensional maps can show chaos only if the map is *noninvertable*. The logistic map is noninvertible because, for a given x_{n+1}, there are *two* possible values of x_n.

2.3.6   The Dripping Handrail

Simple nonlinear models are not limited to those describing a system with a few variables that change through time; such models can also describe spatial patterns. An example of a two-dimensional (spatio-temporal) model is the "dripping handrail" model of Crutchfield and

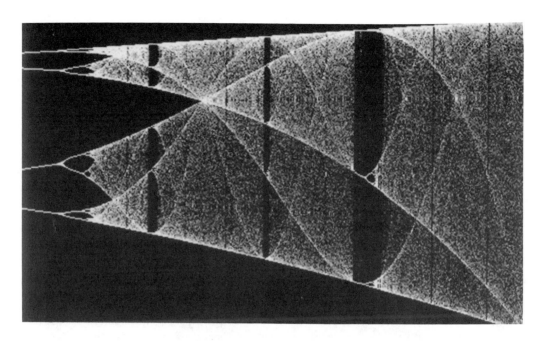

Figure 2.6: The logistic map, in the range $3.5 < r < 4.0$. The gray-scale intensity is proportional to the number of recurrences at each pixel-point.

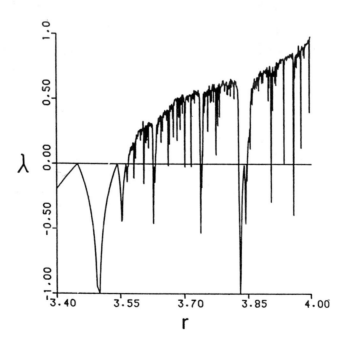

Figure 2.7: The Lyapunov exponent for the logistic map in the range $3.4 < r < 4$. From Wolf, in Holden (1986, p.276).

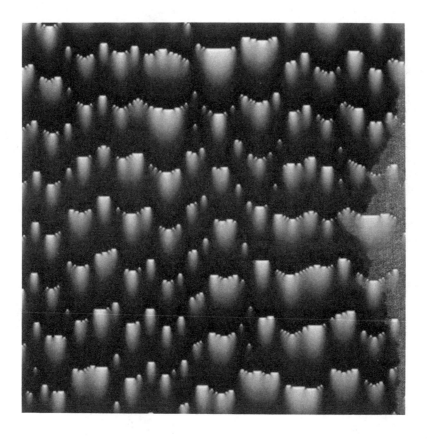

Figure 2.8: Spatio-temporal chaotic pattern created using the dripping handrail, coupled-lattice, model of Crutchfield and Kaneko (1987). The 256 lattice sites are represented left to right across the image and 256 steps through time are represented from top to bottom. The value shown at each location in the image represents the local thickness of water. This example was computed beginning with random initial conditions and performing several thousand iterations before recording the resulting structure.

Kaneko (1987, 1988). This model is a simplified description of the dynamics of a film of water on a handrail (Figure 2.8). Each vertical column in the image represents conditions at a different site s along the handrail; each horizontal row represents one step through time t. Rain falls at rate r on the handrail, which causes the thickness x of the film of water at each site to increase through time (top-to-bottom in the image). When surface tension is exceeded, water drips from that site of the handrail. Coupling between adjacent sites causes water to flow along the one-dimensional lattice that represents the handrail (from sites with a thick film of water to adjacent sites where the film is thinner). This rule can be expressed by

$$x(t, s) = c(x_{t-1,s-1} + x_{t-1,s} + x_{t-1,s+1}) + r$$
$$(2.16)$$

where c is a constant. When x becomes so large that the surface tension threshold for dripping is exceeded, the calculated value of x is reduced by the amount representing a drop of water.

At each point in a two-dimensional image, pixel intensity (or a contour-map value) can be used to represent the local thickness of the film of water. These rules produce a pattern in which the value at any site is a deterministic nonlinear function of 3 values at the preceding time (one at the same site and one at each of the two adjacent sites), in essentially the same manner that each value in a deterministic time series is a function of preceding values.

Crutchfield and Kaneko (1988) noted that the dripping handrail model is attracted toward a periodic state, but that the time to stabilize increases hyperexponentially with lattice size. They presented a computational example with a moderately high number of degrees of freedom (128). This system eventually stabilizes to become periodic, but the time to attain periodicity is extremely long (10^{40} years, if iterations are performed at the rate of 10^{15} per second). For practical purposes the system is nonperiodic, despite the fact that it theoretically would become periodic given enough time.

This example is particularly relevant to the problem of fluid turbulence, where the number of degrees of freedom can be considerably greater than the value of 128 used in these computations. According to Frisch and Orszag (1990) the number of degrees of freedom of a turbulent fluid is given by $R^{9/4}$ per unit volume L^3, where R is the Reynolds number, and L is the length scale used to calculate R. For flows that transport sediment, the calculated number of degrees of freedom can exceed 10^8. Although this relation may define the potential number of degrees of freedom in a fluid, the fluid may not actively employ all of them (Gershenfeld and Weigend, 1994). Nevertheless, this model serves as a warning that fluid flows may take so long to stabilize that experiments may be unable to demonstrate whether or not observed nonperiodic behavior is merely transient nonperiodicity in a high-dimensional system that would eventually become periodic.

2.4 SELF-ORGANIZATION

In the context of chemical systems, this term was popularized by Nicolis and Prigogine (1977). Strangely, though the term appears in the title of their books, it is nowhere given a succinct definition. Perhaps their closest approach is the statement (p. 5):

> ... self-organization in nonequilibrium systems [is] characterized by the appearance of dissipative structures through the amplification of appropriate fluctuations.

Paradigms include purely physical systems, such as thermal convection in fluids, and oscillating chemical systems such as the Belousov-Zhabotinski reaction, and Liesegang phenomena.

Ortoleva et al. (1987) have given a general discussion of self-organization in geochemical reaction-transport systems, and give the following definition (p. 980):

Self-organization is the autonomous passage of a system from an unpatterned to a patterned state without the intervention of an external template.

They point out two necessary conditions: that the system be far from equilibrium (therefore controlled by nonlinear processes) and that "at least two processes active in the system be coupled" (p. 980).

The study of self-organization therefore overlaps to a large extent with the study of nonlinear dynamics, though it concentrates on the appearance of pattern, rather than on chaotic phenomena. It has been applied not only to chemical systems but to many other types. For example, Kauffman (1993) discusses the application to biological evolution, and various applications in the earth sciences have been described in a symposium edited by Ortoleva (1990).

A classic example of self-organization, often cited by Prigogine and coworkers, is the spontaneous appearance of patterned convection cells in Rayleigh–Bénard convection. Other examples include patterned ground in permafrost areas, bedforms, alternating bars in rivers, beach cusps (Werner and Fink, 1993), stylolites (Merino, 1992), and various fractal mineral patterns produced during the crystalization of igneous rocks. It has been suggested by Rinaldo et al. (1993), and Rigon et al. (1994) that river networks, and indeed the entire landscape can be seen as an example of self-organization.

A special type of self-organization has been identified by Per Bak, and called **self-organized criticality** (Bak and Chen, 1991; Bak and Paczuski, 1993; Bak and Creutz, 1994). The fact that fractal distributions have no characteristic scale suggested to Bak that they indicate a system maintained in a critical state. It is known that second-order phase transitions, which occur at a critical temperature, such as the Curie point of a magnetic system, or the superconducting transition temperature, are characterized by the presence of all possible scales

of phenomena (e.g., all possible sizes of ordered domains). Similarly, the critical state in percolation theory is characterized by all possible sizes of interconnected clusters of pore spaces. Bak has suggested, therefore, that if all possible scales are continuously present, the system is probably in a critical state, and must be maintained in this state by the operation of the system itself (i.e., by self-organization).

> ...many composite systems naturally evolve to a critical state in which a minor event starts a chain reaction that can affect any number of elements in the system ...the mechanism that leads to minor events is the same one that leads to major events ...[Such] composite systems never reach equilibrium but instead evolve from one metastable state to the next. (Bak and Chen, 1991.)

Bak and coworkers have applied these ideas to sandpiles, which can be built up to a critical slope, determined by the angle of repose. Addition of further sand produces avalanches whose size distribution (according to Bak) is a power law (i.e., fractal). The timing and size of individual avalanches is unpredictable, indicating that the system as a whole is always maintained close to the critical condition by the steady addition of more sand. Bak has investigated this system using cellular automata (see below), and has applied his idea to cellular automata models of several other phenomena, including earthquakes and forest-fires.

Not only are the spatial properties of self-organized critical systems fractal (in that there is no defining length scale), but time series of events show $1/f$ spectra. These systems differ from the usual (low dimensional) nonlinear dynamical systems, in that they are spatially extended, with many degrees of spatial freedom. Bak has argued that these properties make them a more likely explanation for complex natural phenomena, such as earthquakes, than low dimensional models, such as the slider-block models of earthquakes proposed by some other authors.

2.5 CELLULAR AUTOMATA

[Cellular automata] are dynamical systems in which space and time are discrete. The states of cells in a regular lattice are updated synchronously according to deterministic local interaction rules. Each cell obeys the same rules, and has a finite (usually small) number of states. (Gutowitz, 1991, p. vii.)

The application to a simple, two-dimensional sand pile may be used as an example (Figure 2.9). The pile is made of square blocks of side d and we consider only the right side. The slope is stable if the vertical height h is nowhere greater than $2d$. Starting with a stable configuration, as shown on the left of the Figure, blocks are added at random near the top of the pile, until an unstable configuration results, as shown on the right.

If the slope is unstable, the top two blocks move right and down. This in turn produces further instabilities, so that two more blocks move, and so on. In the example shown, adding one block leads to an avalanch that removes four blocks.

Cellular automata have been applied to many problems. A useful introduction, with simple computer programs, is given by Gould and Tobochnik (1988). Hayes (1984) provides an elementary introduction: see Toffoli and Margolis (1987) or Gutowitz (1991) for a more advanced treatment. Cellular automata have been applied for many years to studies of landscape evolution: recent examples include Smith (1991) and Beaumont et al. (1992).

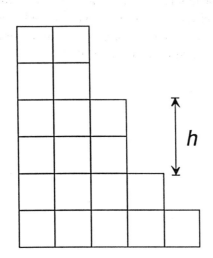 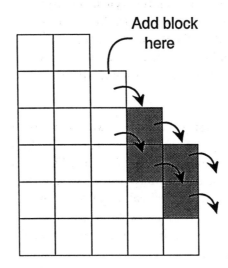

Figure 2.9: Simple cellular automaton for a two dimensional sandpile. The slope is stable only if h is less than 3 blocks; if the slope is unstable, the top two blocks move to the right and down. In the example shown, adding a block makes the slope unstable; after two blocks have moved, the slope further right is unstable, and so on.

Chapter 3

Time Series Analysis I

Roy E. Plotnick
Department of Geological Sciences
University of Illinois
Chicago IL 60680 U.S.A.

Karen L. Prestegaard
Department of Geology
University of Maryland
College Park MD 20742 U.S.A.

3.1 INTRODUCTION

As discussed in Chapter 1, time series whose power spectra exhibit $1/f$ dependency ($1/f$, flicker, or pink noise) have long been known to be ubiquitous in natural and experimental systems (Mandelbrot and Wallis, 1968, 1969a,. b; Schroeder, 1991). For example, this behavior has been recognized in turbulent velocity fluctuation as measured in laboratory flumes (Nordin et al., 1972; Mollo-Christensen, 1973; Nowell, 1978). Mandelbrot and Wallis (1968, 1969a, b) demonstrated that fractional Brownian motions and discrete fractional Gaussian noises provide mathematical models for these $1/f$ noises (see Chapter 1). In addition, the last few years have witnessed renewed attempts to produce general physical models, such as self-organized criticality (SOC), to explain this behavior (Bak and Chen, 1989; Chapter 2, this volume). In addition to spectral analysis, other techniques, such as rescaled range, autocorrelation, and geostatistics have been applied to the analysis of fractal series. In this chapter we describe these methods and illustrate their use in the analysis of real and synthetic data sets with fractal structure.

3.1.1 Autocorrelation: The Persistence of Memory

The calculation of the autocorrelation as a method of studying time series is well established in the geological sciences (Davis, 1986). As discussed in Chapter 1, a key characteristic of fractional Gaussian noises is serial correlation, which produces persistence. Mandelbrot (1971, see also Beran, 1992) defined fractional Gaussian noises as Gaussian random processes having the autocovariance:

$$C(d, H) = \sigma^2[0.5(|d+1|^{2H} - 2|d|^{2H} + |d-1|^{2H})]$$
$$(3.1)$$

where d is the lag, σ^2 is the variance of the series, and H is an exponent that can vary from 0 to 1. If this value is divided by the variance of the series, or if the series is standardized, then this equation can be used to describe the autocorrelation.

Autocorrelograms for a white noise, a persistent noise ($H = 0.8$), and an antipersistent noise ($H = 0.2$) are shown in Figure 3.1. Also shown is the theoretical autocorrelogram, based on Equation 3.4, for an H value of 0.7. For a white noise $H = 0.5$ and $C(d, H)$ thus equals 0; the increments are uncorrelated. When $H > 0.5$, however, $C(d, H)$ only asymptotically ap-

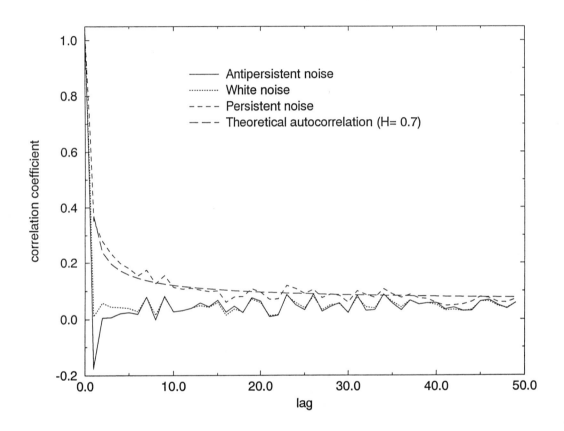

Figure 3.1: Autocorrelograms for white, persistent, and antipersistent noises.

proaches 0 from above, whereas it approaches it from below (negative correlations) for $H < 0.5$. $H > 0.5$ thus represents persistence, while $H < 0.5$ characterize antipersistence. Note that both persistence and antipersistence imply very long memories in the system; e.g., long range serial correlation.

Brownian and fractional Brownian motions are non-stationary and, because they are Markov processes, show a very slow decline of the autocorrelation with the lag (Figure 3.2). Notice that the decay of the correlation is roughly linear and is slower for persistent motions than for Brownian motions and more rapid for antipersistent motions (see next section). These series can be analyzed by converting them to their corresponding noises and examining the autocorrelation of the noises. If the series can be represented as a fractional Brownian motion, then its increments will be a fractional Gaussian noise with corresponding long-range correlations.

3.2 SEMIVARIANCE

For the past decade or so, a popular set of techniques for the description of spatial patterns in geology has been geostatistics (e.g., Davis, 1986; LaPointe and Barton, 1995). One particular method, the production of semivariograms, has been applied to the analysis of fractal data sets (Carr and Benzer, 1991; Carr, 1995; Korvin, 1992).

The semivariance measures the relationship between sets of points a lag s apart. For a set of values n units long, if x_i is the value at one point, and x_{i+s} is the value at a point s units away, then the semivariance γ_s is:

$$\gamma_s = \sum_{i=1}^{n-s} \frac{(x_i - x_{i+s})^2}{2(n-s)} \qquad (3.2)$$

Mandelbrot (1983) pointed out that for frac-

tional Brownian motions,

$$\sum_{i=1}^{n-s} \frac{(x_i - x_{i+s})^2}{(n-s)} \propto |s|^{2H}. \qquad (3.3)$$

As a result:

$$\gamma_s \propto \frac{1}{2}|s|^{2H} \qquad (3.4)$$

(Korvin, 1992).

Figure 3.3 illustrates the semivariograms for the Brownian and fractional Brownian motions on log-log coordinates. As would be expected from Equation 3.4, the slopes of the lines are approximately $2H$ (Carr and Benzer, 1991; Carr, 1995). An important result is that the higher the H value, the larger the expected semivariance; i.e., the larger the spread of the differences at a particular lag. As we will discuss in the next section, this is direct result of the persistence of the increments.

Figure 3.4 shows the semivariance for the corresponding white and fractional Gaussian noises. A comparison with the corresponding autocorrelograms (Figure 3.2) shows these functions to be mirror images. This is the expected result for a stationary, standardized variable such as a noise (Davis, 1986).

3.3 RESCALED RANGE ANALYSIS

The origin of the concepts of fractional noises and motions lies, to a large extent, in Mandelbrot and Wallis' (Mandelbrot and Wallis, 1968, 1969a, b; Mandelbrot, 1995; see also Feder, 1988) efforts to explain some puzzling results of Hurst (1951). Hurst was involved in the design of dams (such as the Aswan High Dam in Egypt) and asked: Given annual variations in rainfall, and thus inflow into a reservoir, how large does a dam have to be in order to produce a constant annual outflow and to not overflow? He began with the assumption that this constant outflow would be equal to the average of the inflow.

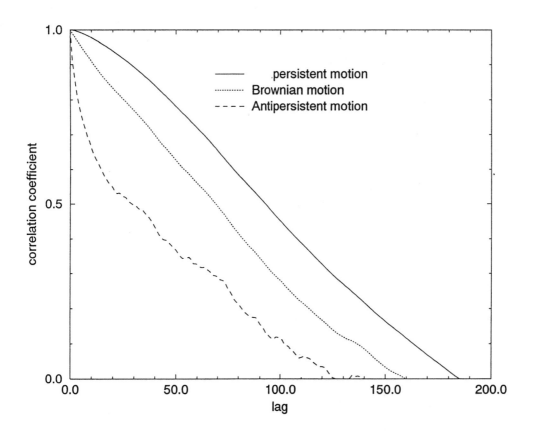

Figure 3.2: Autocorrelograms for Brownian, persistent, and antipersistent motions.

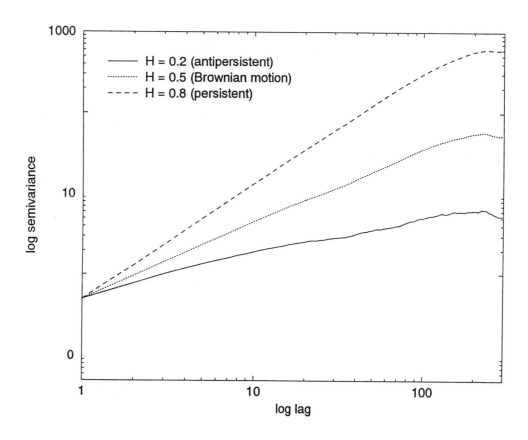

Figure 3.3: Semivariograms for Brownian, perisistent, and antipersistent motions.

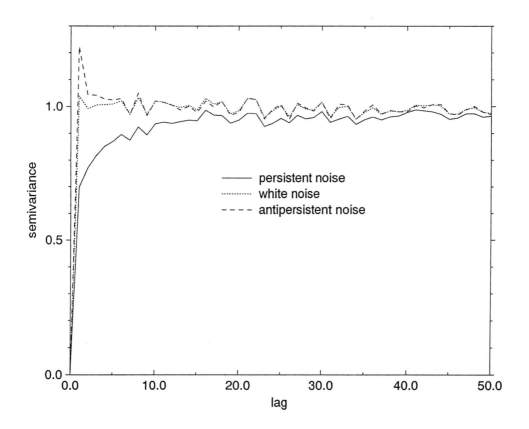

Figure 3.4: Semivariograms for white, persistent, and antipersistent noises.

We can represent this mathematically by designating the annual inflow during year t as $x(t)$. For a period of time of length T the average inflow $\bar{x}$, and thus the outflow, is $\sum x(t)/T$. The annual deviation from the average inflow is thus $x(t) - \bar{x}$.

For simplicity, assume that the variation of annual inflow is a white noise; i.e. that annual rainfall amounts are independent and follow a Gaussian distribution (Chapter 1). Figure 3.5A shows a 128 year synthetic rainfall record. The annual deviations from the average rainfall are shown in Figure 3.5B.

Now consider the record of water level in the reservoir during the entire interval T. In relatively dry years, the inflow will be less than the outflow and the reservoir level will drop. In relatively wet years, the inflow will exceed the outflow and the level in the reservoir will rise. The level of water in the reservoir at any given year $u \leq T$ is thus the accumulated departure $D(u)$ up to time u of the annual inflow from the mean inflow:

$$D(u) = \sum_{t=1}^{u} [x(t) - \bar{x}]. \qquad (3.5)$$

The series of summed deviations for the synthetic record is shown in Figure 3.5C. A little thought will show that since this is the cumulative sum of a white noise, this sequence is a Brownian motion.

During the interval T the value of $D(u)$, and thus the level of water in the reservoir, will at some point reach a maximum level $D(u)_{max}$. At some other time, the level of water, will drop to a minimum value $D(u)_{min}$. The range R of water level during time interval T is thus:

$$R = D(u)_{max} - D(u)_{min} \qquad (3.6)$$

(Figure 3.5C). The range is equal to the minimum necessary size for the reservoir.

R should be proportional to such factors as the size of the drainage basin. To allow comparisons among different systems, Hurst thus

standardized his analyses by dividing the range R by the sample standard deviation S of the series $x(t)$ over the interval T. The resulting ratio R/S is known as the *rescaled range* (Feder, 1988).

It is obvious from examination of Figure 3.5C, that the water level over a short span of years cannot vary too much; the range is necessarily relatively low. As the span of time considered increases, so should the range. In other words, R/S should be proportional to T. Notice that this is roughly equivalent to saying (Section 3.3) that the semivariances for short sequences must be lower than for longer ones.

Mandelbrot and Wallis (1969) examined the dependence of R/S on T by producing a "pox plot" (Figure 3.6). R/S is initially determined for the entire sequence. The sequence is then broken into progressively shorter non-overlapping series of subsequences, where the length T of each subsequence is the lag. The R/S value for each subsequence is then determined and the average R/S for that lag is calculated. R/S is then plotted on log-log coordinates and a line is then drawn or regressed through the average values.

Figure 3.6 shows that, for a Brownian motion, the slope of the regression of R/S on the lag is about 0.5. In other words, for a process whose increments are random and independent (a white noise),

$$R/S \propto T^J, \qquad (3.7)$$

where $J \approx 0.5$. J is known as the Hurst exponent (written as H in some publications). This result essentially mimics the expected result for a diffusion process, which is, of course, modeled as a Brownian motion.

Hurst (1951) and Mandelbrot and Wallis (1969a) applied rescaled range (R/S) analysis to numerous data sequences of geological and geophysical interest and discovered that nearly all J values are greater than 0.6, with most in the range of 0.7. More recently, numerous studies of electric well-logs have shown similar re-

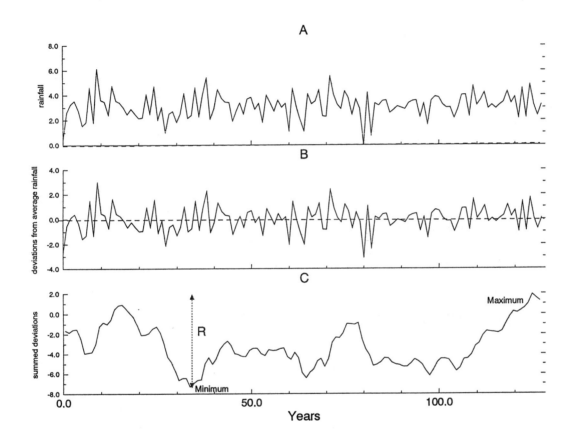

Figure 3.5: Determination of the range for a time series. A. Synthetic rainfall series; a white noise. B. Deviations of the values in series from the overall average. C. Cumulative sum of the deviations; a Brownian motion. R is the range between minimum and maximum values of this series.

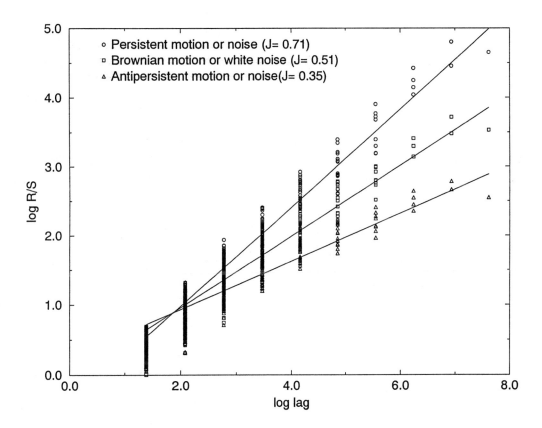

Figure 3.6: Rescaled range analysis (pox plots) for white, persistent, and antipersistent noises and motions.

sults (Hewett, 1986; Emanuel et al., 1987; Aasum et al., 1991; Tubman and Crane, 1995). Fluegemann and Snow (1989) found an J value of 0.78 for a δO^{18} record from a Pacific deep-sea core. Similarly, Hsui et al. (1993) found J values of 0.8 and 0.86 for two Quaternary sea-level curves and 0.92 and 0.96 for long-term sea-level curves for the Cenozoic and Mesozoic-Cenozoic, respectively. Turcotte (1994) applied R/S analysis to analyze flood statistics and found in range of 0.67 to 0.72. The ubiquitousness of J values in excess of 0.5, the expected value from a random process, has been called the "Hurst phenomenon" or "Hurst effect" (e.g., Wallis and Matalas, 1970; Klemes, 1974; Beran, 1992).

Mandelbrot and Wallis (1969a,b) proposed that fractional Gaussian noises and corresponding fractional Brownian motions (Chapter 1) might provide a numerical model for the Hurst effect. Rescaled ranges analysis of a persistent fractional Brownian motion yields $J > 0.5$ (Figure 3.6). This should not be surprising; persistence implies that negative intervals follow negative intervals and positive intervals follow positive. As a result, trends tend to persist and the overall range should be large. In contrast, antipersistence implies that positive increments follow negative and vice versa. There thus cannot be long-term trends, and the range will be small. The values of the Hurst statistic for an antipersistence motion is $J < 0.5$ (Figure 3.2). It should be pointed out that the J value is the same for the motion or its increments, since they are complementary.

Rescaled range analysis has several major advantages and disadvantages. First of all, it is extremely robust. For example, although the examples above are based on Gaussian distributions of the increments, the results are essentially identical for distributions that are skewed, leptokurtic, or platykurtic (Mandelbrot and Wallis, 1969b).

It is also highly robust against the presence of trends. Figure 3.7 shows two sequences; the lower sequence is a Brownian motion whereas the upper sequence is the same motion with a large linear trend added. Both sequences yield J values of 0.5. Numerical experiments with both persistent and anti-persistent noises yield similar results; what R/S is measuring is persistence of deviations from the overall trend. This suggests that it might have utility in studies of evolutionary change over time.

Notice that J and H are closely related; theoretically they are identical. As a result, J has been used to estimator of H. Unfortunately, have shown that estimation using J can yield quite biased results (Wallis and Matalas, 1970; Bassingthwaighte and Raymond, 1994). Values of J above 0.7 tend to be underestimates, while those below 0.7 are overestimates. These effects are exacerbated at low sample sizes.

Finally, it needs to be stressed that although fractional noises and motions are one possible mathematical model for the Hurst phenomenon, a number of other mathematical models have been suggested for this phenomena (see Beran, 1992 and Mesa and Proveda, 1993 for recent reviews). Klemes (1974) and Mesa and Poveda (1993) are especially critical of the "infinite memory" assumptions of the fractal models; i.e., it is hard to imagine a physical model that would be able to produce the asymptotic decline of autocorrelation seen in fractional Gaussian noises. They suggest that non-stationary processes could produce comparable results.

3.4 H AND β

Autocorrelation and Fourier analysis are closely linked mathematically; the Fourier transform of autocorrelation is the power spectrum (Davis, 1986). Similarly, the parameter H in autocorrelation is closely related to the value of β obtained in the periodogram (Mandelbrot, 1983; Voss, 1988).

For Gaussian noises, $\beta = 2H - 1$, so that $-1 < \beta < 1$. Thus in the special case of white noise $H = 0.5$ and $\beta = 0$. Similarly, for motions, $\beta = 2H + 1$, so $1 < \beta < 3$. Thus in the case

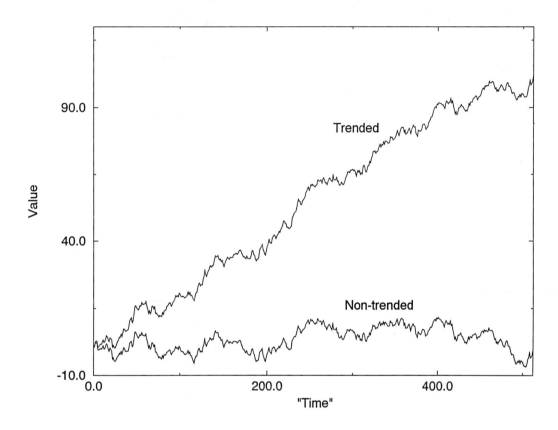

Figure 3.7: Trended and non-trended Brownian motions.

of Brownian motion, where H again equals 0.5, $\beta = 2$. Note that although the H values are the same, the β value for a motion is equal to the value of the corresponding noise $+2$.

3.5 H AND D: ZEROSETS AND LÉVY DUSTS

As discussed in Chapter 1, fractal motions and noises are self-affine rather than self-similar. As a result, it is inappropriate to define their fractal dimension. The fractal dimension can be defined, however, for the zeroset of the motion (Mandelbrot, 1983; Voss, 1988). The zeroset (more generally, isoset) is the set of points defined by the intersection of a motion, such as a Brownian motion, with a fixed value such as zero. Figure 3.8 illustrates the zeroset for a fractional Brownian motion of $H = 0.2$.

It turns out that the points in an isoset form a Lévy dust and thus obey the "uniformly self-similar law of probability." The relationship between D for the Lévy dust and H for the motion is $D = 1 - H$ (Mandelbrot 1983; Voss 1988). For Brownian motions, therefore, the D of the Lévy dust equals 0.5. When H approaches one, the fractal dimension of the dust approaches zero. Due to persistence, a particular value is reached less often. In contrast, when H approaches zero, antipersistence indicates that a particular value is revisited more often. The value of D approaches one, showing that the dust more completely covers the line.

The concept of the zeroset suggests one reason why fractal distributions might occur in stratigraphy. For example, assume that a particular organism lives only at a specific depth below sea level. Further assume, as is the case for several models of sea-level change (e.g., Sadler and Straus, 1990; Korvin, 1992) that sea-level at a locality can be approximated as a Brownian motion. As a result, the timing of the occurrence of the organism at that locality (and thus in the potential fossil record) would

be distributed as a Lévy dust.

3.6 EXAMPLES

3.6.1 Turbulence in a natural stream

Flow velocity measurements were made in East Rosebud Creek, a snowmelt-fed, gravel bed river in southeast Montana, during the summers of 1989 and 1990. The channel bed consists of particles with average diameters ranging from 2.0 to 10 cm in different parts of the channel. Velocity fluctuations were measured with a Marsh-McBirney electromagnetic current meter; the analog signal was digitized and stored on a laptop computer in the field. The probe was located 5 cm above the bed, which was 10%–20% of the flow depth.

A typical time series of velocity fluctuation is shown in Figure 3.9. The data represent 31470 separate downstream velocity measurements taken over 1920 seconds (approximately 16.4 Hz). At first glance, the total series closely resembles a white noise. A closer look at 60 s and 5 s long portions of the series (Figure 3.10), however, clearly shows short-term behavior that more closely resembles that of a motion.

A fast-Fourier transform was used to analyze the first 16,384 points of the series ($=1000$ s) shown in Figure 3.9. The log-log power spectrum is shown in Figure 3.11. The spectrum clearly shows two distinct regions, with the break occurring at a frequencies corresponding to time intervals on the order of 1-5 sec. Regressions of the raw power spectra on frequency yield a β-value of 0.23 for frequencies less than 0.03 (3.3 s) and $\beta = 2.3$ for higher frequencies. These results are consistent with a "pinkish" noise for lower frequencies ($\beta < 1$, so $H = (\beta+1)/2 \approx 0.62$) and a "blackish" noise for high frequencies ($\beta > 1$, so $H = (\beta - 1)/2 \approx 0.65$) and agree with the visual difference between the patterns seen in Figures 3.9 and 3.10.

The semivariogram and the autocorrelation (Figures 3.13 and 3.13) both show patterns con-

Continued on p. 64

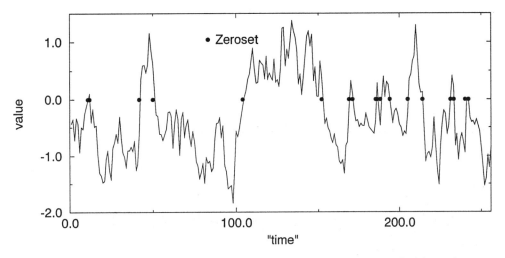

Figure 3.8: Zeroset of a fractional Brownian motion ($x(t) = 0$).

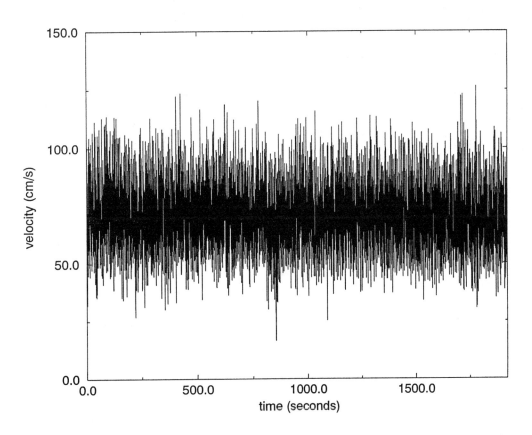

Figure 3.9: Velocity time series in a gravel bed stream, East Rosebud River, Montana, June 20, 1990. 33 minutes long, 31470 data points.

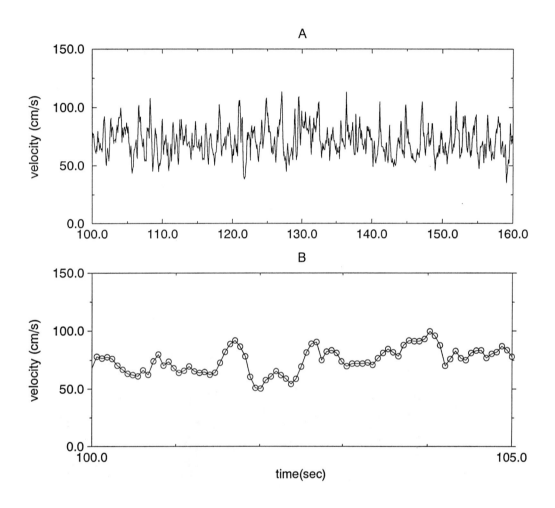

Figure 3.10: Close-ups of sequence in Figure 3.9. A. One minute. B. 5 seconds.

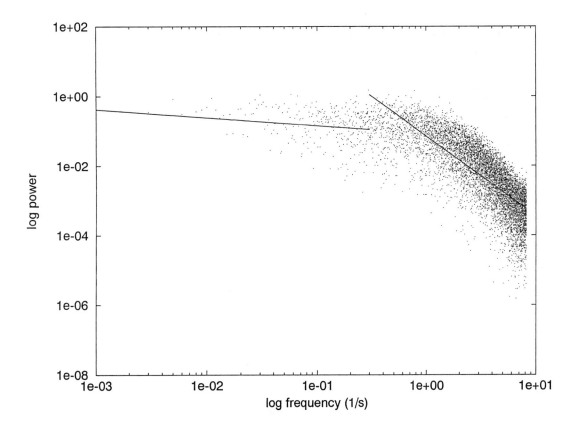

Figure 3.11: Fourier spectral analysis of first 1000 s series in Figure 3.9. Slopes are regressions for $t > 3.3$ s and $t < 3.3$ s.

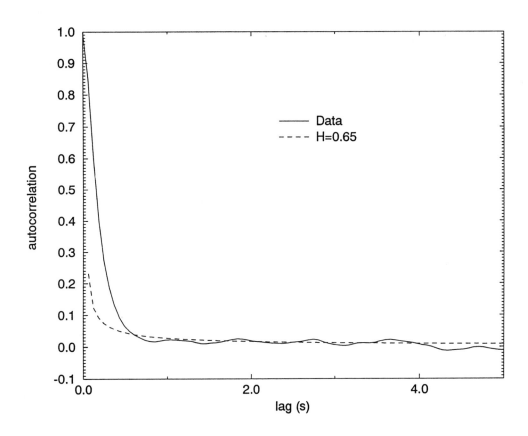

Figure 3.12: Autocorrelogram for lags < 5 s (80 data points) for series in Figure 3.9.

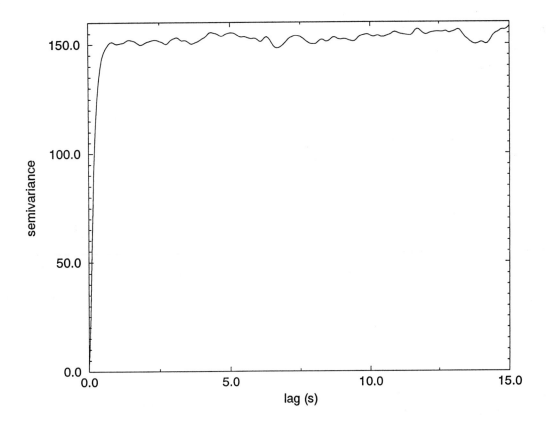

Figure 3.13: Semivariogram for lags < 15 s for series in Figure 3.9.

sistent with rapid decline of the memory of the process up to about 1 s, then a slow approach to random expectation. The theoretical correlogram for an $H = 0.65$ noise is also shown on Figure 3.12. It fits the observed pattern well for time periods beyond 1 s. For shorter lags, the drop in the autocorrelation is slower than expected for a pure noise.

A pox-plot of the rescaled range is shown in Figure 3.14. For comparison, we have plotted the regression for all values of R/S and for the lag averages. The J value estimated from the latter regression is 0.64 ($r = 0.998$).

The results from all three analysis techniques are consistent with a noise with moderate persistence ($H \approx 0.65$) for time periods greater than about 1 sec. The behavior over less than one second more closely resemble that of a motion.

Laboratory investigations have demonstrated the existence of turbulent burst phenomena occurring within a frequency range of 3–8 s (Jackson, 1976). In addition, flume studies of velocity–time series show highly intermittent patterns; this may result from a hierarchy of processes that propagate through the flow at different flow velocities producing non-linear interactions (Mollo-Christensen, 1973; Nowell, 1978). Similarly, $1/f$ noise in turbulence time series data has been explained by fractal and multifractal models of energy dissipation by self-similar eddies (Mandelbrot, 1974; Meneveau and Sreenivasan, 1987).

We suggest, therefore, that the high frequency $1/f$ "black noise" behavior may be due to self-similar eddy cascades, that in this case are constrained by shallow flow depth. The "pinkish" $1/f$ behavior at lower frequencies may be associated with turbulent bursting phenomena that initiate the cascades (Prestegaard and Plotnick, in preparation).

An alternative explanation for the difference in behavior between low and high frequency may be equipment-related. Marsh McBirney e-m current meters commonly filter out high

frequencies (Rubin, personal communication). This is partly because of electrical filtering and partly because of the large volume of fluid that the sensor samples (a single "parcel" of fluid may take tenths of a second to pass through the measurement volume). The observed pattern at high frequencies may thus be due to high speed sampling of a time-averaged input.

3.6.2 Gamma-Ray Logs

As discussed in Chapter 1, these techniques have also been applied to numerous electric well logs. For example, a gamma ray log described in Chapter 1 exhibited a $1/f$ spectrum with $\beta \approx 0.56$ (Figure 1.23), corresponding to a noise with an H of about 0.78. A rescaled range analysis of the same data set yields a J value of 0.81. The autocorrelogram for this sequence in shown in Figure 3.15; along with theoretical autocorrelation for a 0.78 fractional noise. The result from the production of a semivariogram is identical. These results suggest that this sequence can be modeled as an $H = 0.8$ fractional Gaussian noise. As suggested by Hewett (1986) these models can then be used to simulate 2-D variation of reservoir properties (see also Aasum et al., 1991; Hardy, 1992; Tubman and Crane, 1995).

3.7 SUMMARY: STEPS IN THE FRACTAL ANALYSIS OF DATA SERIES

When faced with a sequence to analyze, how does one go about determining if it can be modeled as a fractal? One possible methodology is shown in Figure 3.16. As in all statistical methods, the choice of methods is highly dependent on the nature of the data in question. Three basic data types are utilized in Figure 3.16. Continuous data includes such information as electric well logs, stream or flume velocity measurements, bed roughness (Robert, 1991), or sea-

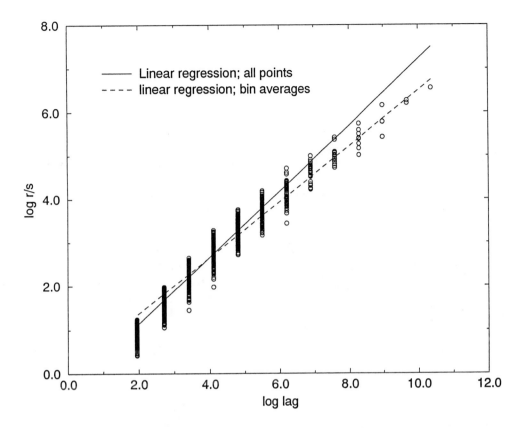

Figure 3.14: Rescaled range analysis of series in Figure 3.9.

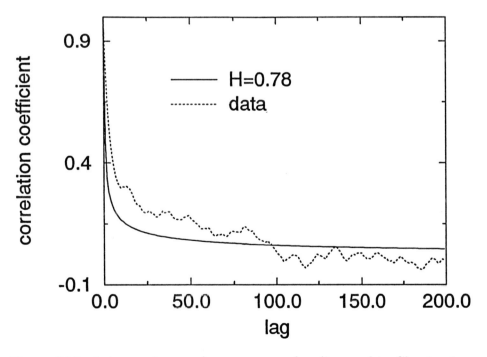

Figure 3.15: Autocorrelogram for gamma ray log discussed in Chapter 1.

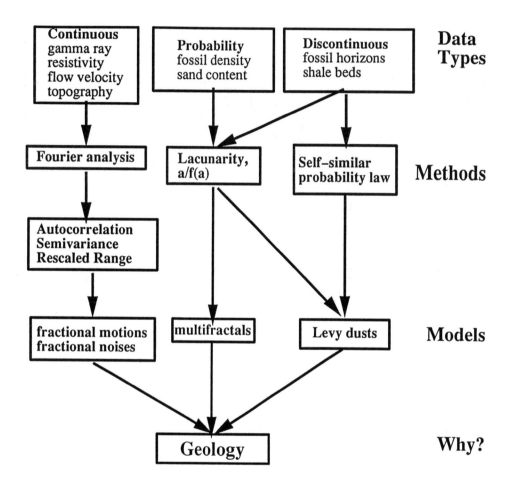

Figure 3.16: Suggested steps in the fractal analysis of data sequences.

floor topography (Malinverno, 1995). "Probability" data basically refers to data series that represent the amount of some material distributed along the sequence. Examples include fossil abundance data, sand percentage, or elemental concentration. Discontinous data refers to the distribution of events in a sequences, such as the occurrence of fossiliferous horizons, beds of a particular lithology, or asteroidal impacts.

The first stage in the analyis of continuous data is a Fourier spectral analysis. This should reveal whether the series shows $1/f$ frequency dependence and the nature of that dependence. The value of suggests whether a fractional noise or motion model is appropriate. This is then followed with some form of lag-domain analysis, such as R/S, autocorrelation, or semivariance. Again, this can be compared with the expected result for a sequence that is a noise or a motion.

Probability data can be analyzed using the various techniques for analyzing mass distributions such as $\alpha - f(\alpha)$, or lacunarity (Chapter 1). These techniques will reveal whether some form of multifractal model (Feder, 1988) is an appropriate description of the data.

Discontinuous events can also be analyzed with lacunarity or with some other method for determing the fractal dimension (such as box counting; Chapter 1) to determine if the events follow a random or fractal distribution. The distribution of gaps between events can also be compared with the "uniformly self-similar law of probability" (Chapter 1) to see if a Lévy dust model is appropriate.

In many ways, the fractal analysis of empirical data is still in its infancy; for example, many techniques do not have associated confidence intervals. In addition, statistical results, such as those produced by rescaled range analysis, may be consistent with several different matehmatical models. Finally, stating that a particular pattern can be modeled as a fractal is equivalent to stating that the Earth can be modeled as a sphere; it does not provide an explanation for the cause of the pattern. Determining the

reason why an object or series is fractal depends upon the use of physically realistic process models.

APPENDIX: A Note on Computer Programs

Computer programs for simulation and analysis of fractal sequences are becoming widely available. A particularly useful source is Russ (1994), which includes fractional noise generators, Fourier and rescaled range analysis, and Richardson analysis. This reference is also the best available discussion of fractal techniques for the analyis and simulation of surfaces. Many statistical packages have provisions for Fourier analysis and autocorrelation. A readily available library of geostatistical programs, including semivariance, is Deutsch and Journel (1992). Fortran and Basic programs for lacunarity and R/S analysis are available by e-mail from Plotnick (`plotnick@uic.edu`). Those wanting to do their own programming should look at the pseudocode algorithms provided by Saupe (1988).

Chapter 4

Time Series Analysis II

Gerard V. Middleton
Department of Geology
McMaster University
Hamilton ON L8S4M1, Canada

4.1 EMBEDDING

Most methods of analysing time series carry out the numerical analysis on the time series itself. This is because the investigator is basically interested in the variable $x = f(t)$ that has actually been measured. But an alternative approach is to suppose that the variable that was measured is merely one of several that might be measured as output from a dynamical system, which might indeed be better characterized using some other variable. For example, fluid convecting in a box might be characterized by a probe measuring temperature or velocity at one point in the box, or by some other measurement, as a function of time. If the governing equations for the system are not known, it is also not possible to identify the fundamental variables which should be measured, or even how many of them there are. In this case, one might think that it is impossible to use measurements made on a single, arbitrarily chosen variable to reconstruct any important properties of the full multidimensional system. As we have seen, it is generally not even clear from simply examining the output x, whether or not the system producing x is stochastic or a nonlinear dynamical system of relatively low dimension.

But if, in fact, the signal is a product of a low dimension deterministic system then it turns out that it is possible to reconstruct all the major topological properties of the system, by a technique known as *embedding*. The basic idea seems to have been discovered independently by several different people or groups, so it is variously known as "the Santa Cruz conjecture" (Packard et al., 1980: for some unusual historical insights about this group see Bass, 1985), or as a theorem named for one or more of Whitney (1936), Takens (1981) or Mañé (1981).

Suppose the original time series consists of values of x measured at regular time intervals Δt. We can designate such a series either as

$$x(t), x(t + \Delta t), x(t + 2\Delta t), \ldots, x(t + (N - 1)\Delta t$$

or as

$$x(1), x(2), x(3), \ldots, x(N)$$

The commonest method is known as the **method of delays**: from the original time series we use a delay m, corresponding to a time interval $T = m\Delta t$ to produce the d-dimensional time series shown in Table 4.1, which may be written more compactly as the series of coordinates of the embedding space

$$x(t), x(t + T), x(t + 2T), \ldots, x(t + (d - 1)T)$$

A suitable choice of lag time can produce trajectories from a single time series which bear a striking resemblance to the trajectories produced by the entire set of equations. This is shown for the Rössler system in Figure 4.1.

$$
\begin{array}{ccccc}
x(1), & x(2), & x(3), & \ldots, & x(N) \\
x(1+m), & x(2+m), & x(3+m), & \ldots, & x(N+m) \\
x(1+2m), & x(2+2m), & x(3+2m), & \ldots, & x(N+2m) \\
\cdots & \cdots & \cdots & \cdots & \cdots \\
x(1+(d-1)m), & x(2+(d-1)m), & x(3+(d-1)m), & \ldots, & x(N+(d-1)m)
\end{array}
$$

Table 4.1: Time series produced by embedding in d-dimensional space.

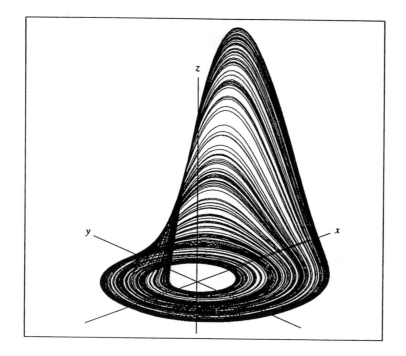

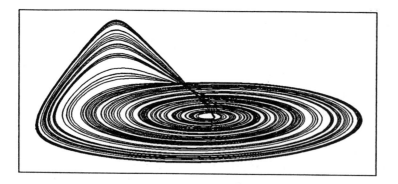

Figure 4.1: The Rössler attractor (top), and its reconstruction from a time series, using the method of delays. From Peitgen et al. (1992, p. 688 and 748).

The Takens theorem states that, under rather general conditions, the topological properties of the original attractor can be reconstructed from an embedding in d_e space, provided that $d_e \geq 2d_a$ where d_a is the topological dimension of the attractor (in general, $n - 1 < d_a < n$, the number of variables $x_1(t), x_2(t), \ldots, x_n(t)$ defining the state space of the dynamical system). The theory of embedding makes use of fairly advanced concepts in topology: it is fully explored in Sauer et al. (1991), and summarized in Ott (1993) and in Chapter 5 of Ott et al. (1994). The reason for the inequality can be understood from the following heuristic reasoning.

First we recall that points have a topological dimension of zero, curves have a dimension of one, and surfaces (in 3-space) have a dimension of two. We expect that two surfaces in three-dimensional space will intersect, and the intersection will be a line (Figure 4.2). There are, of course, a few exceptions, such as two parallel planes. Similarly, a curve commonly intersects a surface at a point, while two curves do not intersect at all. In general

$$d_i = d_1 + d_2 - d_s$$

where d_i is the dimension of the intersection, d_s is the dimension of the space, and d_1 and d_2 are dimensions of the manifolds (curves or surfaces) in that space. For dynamical systems, trajectories in state space never intersect, since the system is (by definition) a deterministic one where a future state depends uniquely on an earlier state. Setting $d_i < 0$ ensures that two manifolds do not intersect, and if we want to exclude the intersection of a manifold with itself we require that

$$d_s > 2d_m$$

The Lorenz attractor has a dimension somewhat larger than 2, so apply this argument we might expect that to be sure that it is properly "unfolded" we need an embedding space of 5. In reality, an embedding in a lower dimension space

(e.g., 3) may be sufficient—but it is not guaranteed by Takens' theorem.

4.2 OPTIMUM DELAY TIME

To achieve a good reconstruction of the topology of the original attractor it is essential to chose the right time delay. This is because if the delay chosen is too small, then the delay variables $x(t + iT)$, $i = 0, 1, 2, \ldots, (d - 1)$ are highly correlated—which means the reconstructed attractor surface will be nearly flat. If the delay is too large, then there is almost no relationship between the variables, and the attractor will appear to be a random cloud of points. So we want something between those extremes.

There is no theory which indicates exactly what delay should be chosen but two guiding rules have been suggested:

- Plot the *autocorrelation function* for the time series. This shows how the linear correlation coefficient r between terms varies as the delay time is increased. Most series show a rapid decline in r followed by an oscillation and/or an asymptotic approach to zero. The rule is to chose the delay corresponding to the first minimum.

An objection to this technique is that we are interested in nonlinear not linear relationships between terms in the time series. So it seems inappropriate to use a linear measure, like the correlation coefficient. One can always argue, however, that nonlinear functions are likely to be almost linear for short delay times. This assumption is, in fact, fundamental to the *local linear approximations* used in Chapter 5 to make nonlinear predictions.

- Instead most workers now favor calculating the *average mutual information* $I(T)$ between measurements made at t and $t + T$

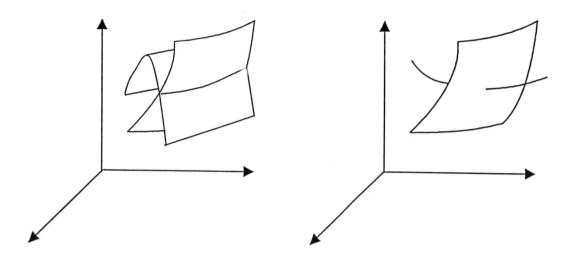

Figure 4.2: Intersection of two surfaces along a curve (left), and intersection of a surface with a curve at a point (right).

(Fraser and Swinney, 1986). This measure is derived from information theory, and given by the equation

$$I(T) = \sum p_{t,t+T} log_2 \left(\frac{p_{t,t+T}}{p_t \cdot p_{t+T}} \right)$$

where p_t and p_{t+T} are the probabilities distributions for $x(t)$ and $x(t+T)$ respectively, and $p_{t,t+T}$ is the joint probability of $x(t)$ and $x(t + T)$. Estimates of these probabilities can be obtained by plotting histograms of values from the observed time series. If the time series is stationary (does not change its average statistical properties with time) then we expect $p_t = p_{t+T}$. The joint probability is estimated from the histogram of joint occurences of $x(t)$ and $x(t + T)$ (Abarbanel, et al., 1993). Once again we choose the time delay corresponding to the first minimum in the plot of I versus T.

Having chosen the best delay time, we now have to determine the appropriate embedding dimension. If the time series is a product of a stochastic process, or of a very high dimensional dy-

namic system, then there really is no appropriate embedding. Realistically, calculations can rarely be carried out on embeddings greater than 10. This is partly because of the large amount of computing that is involved (with its attendant possibility of numerical errors) but mostly because observations made on natural or experimental systems are not error-free, and the presence of error in the measurements generally cannot be separated from the presence of more variables (higher dimensions) in the system.

4.3 FALSE NEIGHBORS

Assume first that a time series is a product of an n-dimensional nonlinear dynamical system with an attractor of dimension $n - 1 < d_a < n$. If the embedding dimension is smaller than the true dimension of the attractor then some trajectories will appear to cross: this means that at least some of the points determines along the "crossing" trajectories appear to be much closer together in space than is really the case in the true state space, or in an embedding dimension that is large enough to reproduce the topology of the original attractor. This suggests the following method to determine the correct embedding dimension (Kennel et al., 1992).

First embed the time series in a low dimension d_e. Choose points at random on the trajectory, and for each point determine its neighbors, (other than those that are close because they are points close in time to the original point). Neighbors, in other words, are points P that are separated in time from the reference point O by some increment Δt, but are closer in the embedding space than some normalized distance

$$\Delta = \frac{1}{d_e} \left(\sum_{i}^{d_e} (x_{oi} - x_{pi})^2 \right)^{1/2}$$

Now increase the embedding dimension to $d_e + 1$ and check to see how many of the original neighbors are still closer than Δ. Those that are no longer neighbors can be called *false neighbors*— in a large enough embedding space they did not lie particularly close to each other. The method originally described by Kennel et al. (1992) is the same in principle: for each point in the time series we determine its (single) nearest neighbor and its distance Δ. If, after increasing the dimension, the distance Δ increases more than a critical amount (10 is suggested by these authors) then the neighbor is classified as a *false nearest neighbor*.

The number of false neighbors is expected to decrease until a large enough embedding space is reached. At this point the number should be zero, or at least very small (allowing for points on separate trajectories that just happen to be very close, or for the presence of error in the data). Increasing the embedding dimension further does not produce any further change in the number of neighbors.

This behavior may be contrasted with what would be observed if the data were stochastic. In this case, the number of neighbors will continue to decrease but will not reach zero no matter how large an embedding dimension is used. Figure 4.3 shows percentages of false neighbors

plotted for a time series generated by the Lorenz system, and for a series of random, uncorrelated data. Kennel et al. (1992) pointed out that the false nearest neighbor criterion alone is not sufficient to distinguish an attractor from random data, because the nearest neighbor for a limited number of random data may not, in fact, be very close—their distance may be comparable with the size of the "attractor" (the complete range of data, in the case of random data). A measure of the size of the "attractor" in this case is given by the total standard deviation of the time series. For random data, the distance between nearest neighbors approaches this value as the embedding dimension is increased—and this can be used as a second criterion to detect false nearest neighbors. For details of the algorithm see Abarbanel et al. (1993).

Most recently, Fredkin and Rice (1995) have shown that the false nearest neighbor method can falsely indicate that a stationary autocorrelated random process is deterministic. They suggest ways to overcome this defect by removing the effect of the autocorrelation. A revised form of false nearest neighbors, called "false strands" is currently under development by M.B. Kennel and H.D.I. Abarbanel (preprint available from the authors at `inls1.ucsd.edu`).

4.4 CORRELATION DIMENSION

Probably the best known technique for analysing embedded time series is the one originally proposed by Grassberger and Procaccia (1983). Indeed, this is the *only* technique described in many texts.

To understand the basis of this technique, consider the two-dimensional examples shown in Figure 4.4. The first diagram (A) shows a simple curved trajectory in state space. Chose a reference point, and then draw circles of radius r around this point. The number of points included in the circle is expected to be propor-

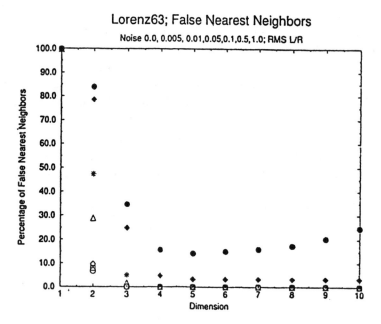

Figure 4.3: Percentage of false nearest neighbors for a time series generated by the Lorenz system (open circles), and for the Lorenz system combined with increasing amounts of uncorrelated random noise. The case where the variance of the noise is as large as that of the signal is shown by the solid circles. From Abarbanel et al.(1993).

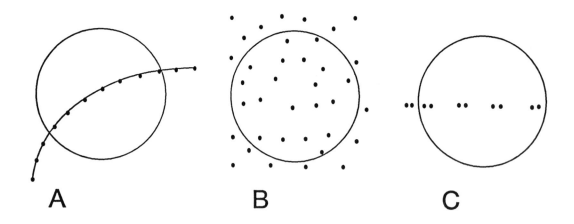

Figure 4.4: Method for calculating correlation dimension, in an embedding space of two dimensions. A. Points lying on a simple curve; B. Random data points; and C. A Cantor set. Based on a diagram in Bergé et al. (1984, p. 150).

tional to the first power of r. The second diagram (B) shows a set of random points. In this case we expect the number of points to be proportional to r^2. The third diagram shows a fractal set of points (the Cantor set). In this case we expect the number of points to be proportional to r^d where $d = 0.63\ldots$ is the fractal dimension of the set.

Now, if the same data were embedded in three dimensions, the curved trajectory would still remain a curve, with a dimension of one, and the number of points in a sphere of radius r would still be proportional to r, the random data would be scattered throughout the space, so the number of points would be proportional to r^3, and the Cantor set would remain a set of points with a fractal dimension of $0.63\ldots$, and the number of points in a sphere would be proportional to $r^{0.63}$. Therefore a plot showing how the number of points $N(r)$ in a neighborhood of size r varies with r provides a technique for determining the dimension of the attractor (if one exists) and a way of distinguishing a low dimensional attractor from random, uncorrelated data.

In practice, we determine $N(r)$ for all points in the embedded time series, and write the averaged value as the *correlation function*

$$C(r) = \lim_{n \to \infty} \frac{1}{n^2} N(r)$$

where $N(r)$ now means the number of pairs of points (from a total number nearly equal to n^2) which are closer than r. This is often written using the concept of a "Heavyside function" H, which is defined by the rule $H(x) = 1$ if x is positive or zero, and $H(x) = 0$ otherwise. Then if the distance between the ith and jth point in space is written as $|x_i - x_j|$

$$C(r) = \lim_{n \to \infty} \frac{1}{n^2} \sum_{i,j}^{n} H(r - |x_i - x_j|)$$

Details of the Grassberger-Procaccia algorithm are given in the book by Parker and Chua (1990).

For large data sets that are relatively free from noise, this method works well, but it becomes inaccurate and unreliable for small, noisy data sets. Limited data often produce plots that suggest the presence of an attractor, even where no real evidence actually exists: large amounts of good quality data are required to determine the presence and dimension of attractors with dimensions greater than three. Exactly how much data are required, and whether or not noisy data can be preprocessed in some way, remain controversial. Theiler (1986) suggested that distortion of the $\log C(r)$ plots at low values can be minimized by excluding nearest neighbors from the plot: the number excluded being determined by trial and error. For other references on this subject see Abarbanel et al. (1993), Ott et al. (1994), and Tsonis (1992).

4.5 LYAPUNOV EXPONENTS

A characteristic feature of chaotic dynamic systems is that two trajectories that start from neighboring points on the attractor diverge from each other exponentially with time. Lyapunov exponents quantify the rate at which this divergence takes place (see Chapter 2 of these notes).

It can be shown that in dissipative systems, a small volume of points, originally of size Δr, will contract in volume as it moves on the surface of the attractor (*Liouville's theorem*, e.g., Thompson and Stewart, 1986, p.221; Ott et al., 1994, p.5). Therefore, if stretching of the volume due to exponential divergence takes place in one direction, contraction must take place in another (orthogonal) direction. Stretching is indicated by a positive Lyapunov exponent and contraction by a negative exponent. In general, if the state (or embedding) space has d dimensions, then there will be d Lyapunov exponents.

One way to detect chaos in a time series is therefore to embed the time series in a d-space, and calculate the Lyapunov exponents. A great variety of numerical techniques have been designed to carry out this calculation, beginning with the work of Wolf et al. (1985). A review and reprints of a few of the more recent papers are contained in the book by Ott et al. (1994). Other reviews can be found in Abarbanel et al. (1993), Grassberger et al. (1991), and Kaplan and Glass (1993). The method originally used by Wolf can lead to inaccurate results and improved methods were suggested by Briggs (1990) and Brown et al. (1991).

The simplest techniques attempt to determine only the largest Lyapunov exponent. One method, which was specially designed for small data sets, was published by Rosenstein et al. (1993), and software programs (`mtrchaos` and `mtrlyap`) that implements this method on the PC are available from the author and at several Internet sites (see Appendix II).

4.5.1   Local vs Global Properties

The properties of strange attractors may vary greatly in different parts of state space. For example, the sensitivity to initial conditions, and therefore the Lyapunov exponents, may vary from one part of the attractor to another. Abarbanel (1992; see also Abarbanel et al., 1993) has described methods to calculate *local* Lyapunov exponents.

The embedding dimension needed to unfold an attractor, as determined by false nearest neighbors, may also be larger in some regions than in others (Abarbanel and Kennel, 1993; Abarbanel et al., 1993).

Some parts of an attractor may be visited much more frequently than others. This has been shown most clearly for maps, such as the Hénon map (Arneodo et al., 1987), and suggests that rather than using a single fractal dimension to characterize an attractor, one might use the attractor as a support for a probability measure, and characterize the probability distribution over the surface by a multifractal distribu-

tion. For a recent review see Muzy et al. (1994).

Local variation in the properties of attractors is also important for the forecasting techniques discussed in Chapter 5.

4.6   USE OF SURROGATE DATA

Almost all nonlinear methods for the analysis of time series have been developed recently, and suffer from the disadvantage that there are no established statistical tests for the significance of the results. The methods are generally tested on output from well-known low dimensional dynamic systems, such as the Lorenz system, and the results are contrasted with those obtained by using random time series (using uniform white, or Gaussian white noise, i.e., uncorrelated noise with a uniform or Gaussian distribution). In some cases, noise has been added to the output from the model system, to see how it affects the results. Though these procedures are valuable, they do not exclude the possibility that similar results might be produced by some time series which is neither statistically random, nor produced by a low dimensional chaotic system.

Two techniques that have been suggested to address this problem are:

1. If evidence has been found for operation of a low-dimensional nonlinear dynamic system, then it should be possible to use this information and a part of the original time series to produce nonlinear forecasts of the remaining part of the time series that are *better* than those that can be produced using the usual linear forecasting techniques. Nonlinear forecasting techniques are described in Chapter 5 of these Notes.

2. The times series may be analysed by spectral analysis: the spectrum shows the contribution to the total variance of each frequency that can be detected in the original

time series. For chaotic data, the spectra are generally "broad band," i.e., they show substantial contributions from a wide range of frequencies, similar to that shown by data from a random, or strongly stochastic process. The spectrum takes no account of the phases of the frequency components of the Fourier series fitted to the original data, but these phases are known, and can be used to reconstruct the original time series from the spectrum. This is how the spectrum is used as a filtering or smoothing device: a part of the spectrum is deleted or modified, and then used to reconstitute a filtered version of the original series (mathematical details are given in any text on spectral analysis, such as Newland (1993), and computer programs are given by Press et al., 1989).

It is also possible to use the spectrum of a time series to develop a *surrogate data set*, consisting of data that has the same spectrum as the original data, but randomized phases. This may produce data sets that, to the eye, closely resemble the original set, but lack any of the nonlinear information contained in it (e.g., Kaplan and Glass, 1993). If this surrogate data is then used as input to determine the embedding space, or the correlation dimension, or the Lyapunov exponents of the embedded series, the results should be quantifiably different from that of the analysis of the original data. If not, there is reason to doubt the results obtained using these techniques (Theiler et al., 1992).

These techniques essentially set up "null hypotheses." The result of applying a test to data resulting from such a null hypothesis are then compared with the result of applying the same test to the real data. As the null hypotheses become increasingly sophisticated, the limitations of the analytical techniques become more evident. Relatively simple techniques can suc-

cessfully distinguish periodic signals, chaos, and uncorrelated noise, but the same techniques may be incapable of identifying correlated noise. Even more advanced techniques are required to distinguish chaos from combinations of randomness and periodicity, such as correlated phase distortion given by

$$x(t) = \sin(t + e), \qquad (4.1)$$

where t is time and e is correlated noise (Rubin, 1992).

The problem of arriving at a unique explanation of observed data is not peculiar to the analysis of time series. By applying certain tests, we can conclude that the results are consistent with the hypothesis that the data were generated by a low-dimensional chaotic system—but we can never be absolutely sure that the same result could not have some other explanation. The same is true of any scientific hypothesis or theory.

Figure 4.5 shows an example described by Elgar and Kadtke (1993). These authors show that oxygen isotope (paleoclimate) data from ODP site 677 suggest the operation of a nonlinear dynamical system with a dimension of about 6. There are reasons to remain skeptical about determination of the dimension of such a high dimensional attractor from such a small amount of data, but their conclusion is greatly strengthened because they show: (i) the effect does not seem to be much affected by smoothing the data; (ii) analysis of the surrogate data shows no sign of such an attractor; (iii) analysis of the supposed forcing mechanism (summer insolation) shows a similar dimension.

4.7 AN EXAMPLE: THE GREAT SALT LAKE

Most of the well known examples of chaotic attractors are either model equations, such as the Lorenz or Rössler systems, or carefully controlled experimental systems such as the mechanical systems described by Moon (1992), the

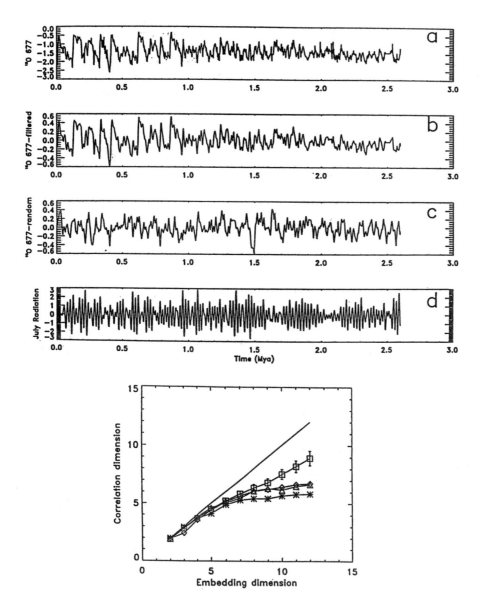

Figure 4.5: Upper diagram shows: (a) Oxygen isotope data (1298 points) from ODP site 677; (b) Same data, but filtered to remove high frequency components; (c) Surrogate series derived from the original data; (d) Theoretical July solar insolation for the same time period. Lower diagram shows the correlation dimension obtained for each of these time series: diamonds are original data; triangles are filtered data; square are surrogate data (with one standard deviation error bars), and stars are insolation data (from Elgar and Kadtke, 1993). The straight line is the result to be expected froma completely random time series.

fluid mechanical systems described by Bergé et al. (1986), the chemical examples described in Field and Györgyi (1993) and Scott (1994), or the biological examples described by Olsen and Degn (1985). A classic example of the reconstruction of an "almost-natural" time series is the analysis of a dripping tap, originally given by Shaw (see Crutchfield et al., 1986) and more recently duplicated by Cahalan et al. (1990). But although Lorenz' work was designed to model the circulation of the atmosphere, attempts to detect low-dimensional dynamical chaos in real meteorological or climatic data have produced results that were controversial, at best. For reviews, see Elgar and Kadtke (1993), Lorenz (1991, 1993), Ruelle (1994), Tsonis (1992) and Zeng et al. (1993). The same can be said for much of the work in ecology, reviewed by Hastings et al. (1993).

Most attempts to detect chaos in geophysical time series have also given results that were negative (Beltrami and Mareschal, 1994), or that were interesting—but inconclusive (Cortini and Barton, 1993; Dubois and Cheminée, 1993; Godano and Salerno, 1993; Mudelsee and Stattegger, 1994; examples in Newman et al., 1994). Possible explanations are that these systems were the product of high-dimensional systems (e.g., see the discussion of Bak's ideas about earthquakes, described in Chapter 2), or that (on the observational time scale) the system was being driven by another system, whose behavior changed significantly with time (i.e., the forcing was non-stationary).

These reservations about progress so far are not meant to discourage further attempts, but to show that documenting chaos in a natural stratigraphic or geophysical time series requires high-quality data and analysis by more than one well-tried method. The report that follows satisfies both these criteria, but not the requirement that surrogate data or other hypotheses should also be tested.

The example to be discussed is described by Abarbanel and Lall (in press), and in several unpublished manuscripts by Lall and coworkers. The level of the Great Salt Lake in Utah has been recorded carefully every 15 days since 1847, and the volumes (in 10^7 acre feet) calculated from these data by T.B. Sangoyomi (unpublished Ph.D. thesis, Dept. Civil and Environmental Engineering, Utah State University, 1993) are shown in Figure 4.6. The power spectrum of these data is shown in Figure 4.7 — it shows several peaks, corresponding mainly to yearly cycles or their harmonics, rising above a background of "broadband noise." After applying several, recently-developed techniques to these data, Abarbanel and Lall have concluded that much of the apparently random variation can be explained as the product of a nonlinear dynamical system with a dimension of four, producing a strange attractor with a dimension of roughly 3.5.

Figure 4.8 shows the average mutual information for this time series. The first minimum is in the region of 12–17 intervals of 15 days. Embedding of the time series for most of the other analyses reported was therefore carried out using a delay of 12 time intervals (180 days, or about 6 months).

Figure 4.9 shows false nearest neighbors calculated globally on the emmbedded data, using the methods described by Abarbanel et al. (1993). It indicates rather clearly that the embedding dimension is 4: at this dimension there are no false nearest neighbors, and none appear at higher dimensions. The result is confirmed by using the method of local false nearest neighbors described by Abarbanel and Kennel (1993).

Figure 4.10 shows the result of applying the Grassberger-Procaccia algorithm to the same data, reported by Lall et al. (in press). The results do not lend themselves to a precise determination of the attractor dimension, but they do indicate that it is about 3.5, consistent with the embedding dimension of 4 determined from false nearest neighbors. Determination of Lyapunov exponents suggests values of about $\lambda_1 = 0.17$, $\lambda_2 = 0$, $\lambda_3 = -0.13$, $\lambda_4 = -0.68$. The pres-

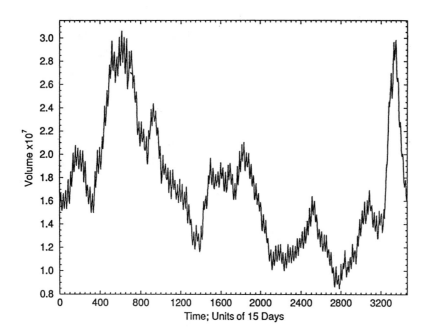

Figure 4.6: Great Salt Lake Volume, in units of 10^7 acre feet, as computed from lake levels measured every 15 days since 1847. From Abarbanel et al., in press.

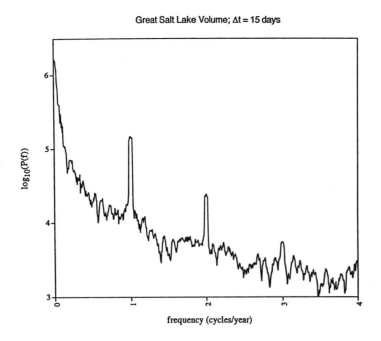

Figure 4.7: Power spectrum of the Great Salt Lake time series. Units on the frequency axis are 1/85 years. From Abarbanel and Lall, in press.

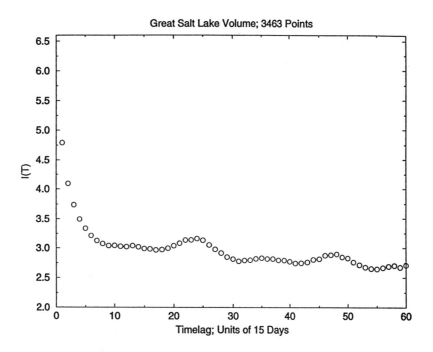

Figure 4.8: Average mutual information for the Great Salt Lake time series.

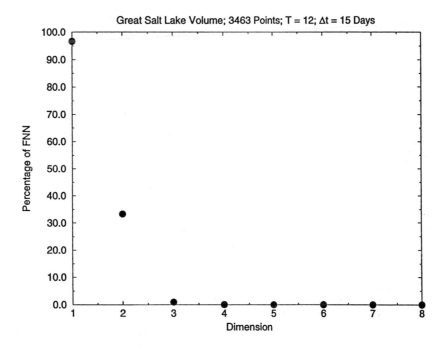

Figure 4.9: Global false nearest neighbors for Great Salt Lake data.

ence of a positive exponent suggests a chaotic system: errors should grow as $\exp(0.17t/\tau)$ where τ in the time interval (15 days) — so the limit of reliable (nonlinear) prediction should be of the order of a few times $t = 15/0.17 \approx 100$ days. This was confirmed by using earlier parts of the time series to predict later parts. Such predictions were very good for the next 15 days, and gave fair results up to 150 days. Determination of the Lyapunov exponents also allowed an estimate of the dimension of the attractor, using the Kaplan-Yorke conjecture: the estimate is a little larger than 3, which is consistent with the other results reported.

This analysis of the data, of course, is not exhaustive, and so it does not demonstrate that some other explanation is not possible. There are some indications that the data are not stationary, which might affect the results. Would the analysis give similar results if the known annual periodicity was removed first? Alternative hypotheses, such as that of correlated $(1/f)$ noise have not yet been adequately explored. A more convincing case could be made by showing that nonlinear models perform better than linear models in predicting the time series, or by comparing the results with those obtained using surrogate data. Neverthless, this is an interesting example of a high-quality geophysical time series, that has been analysed by workers conversant with state-of-the-art nonlinear dynamical techniques.

4.8 CONCLUSIONS

Given a possibly chaotic time series, the following is a list of steps which might lead to a more-or-less complete analysis, in the present state-of-the-art:

1. Check for trends (non-stationarity). If there are any, they must be removed (perhaps by regression) before further analysis. Of course, apparent non-stationarity may simply mean that the time series is too short to permit effective analysis: in the context of chaotic data, this means the observed trajectories do not adequately cover the entire attractor.

2. Calculate the autocorrelation function, and perform a spectral analysis. These will give a general idea what kind of series is involved, and whether there are any major cyclicities.

3. If the series is long enough (more than 1000 data), calculate the average mutual information. Use the first minimum of this, or the autocorrelation function to estimate the best time delay.

4. Embed the series in embedding spaces (up to about 10, depending on the number of data), and use the false nearest neighbor method to estimate the smallest global embedding dimension.

5. If the embedding dimension is less than 4, try visualizing the trajectories using two dimensional plots.

6. Determine the attractor dimension using the Grassberger-Procaccia algorithm.

7. Determine the largest Lyapunov exponent, or the complete set of exponents if the data set is large enough. If determination of the complete set seems to give good results, use it to estimate the attractor dimension.

8. Use the calculated spectrum to generate one or more sets of surrogate data: rerun all tests (items $\geq$ 4 above) on the surrogate data sets.

9. Use prediction techniques to determine whether a better prediction of the later part of the time series is possible using nonlinear than using linear techniques (see Chapter 5).

These are, of course, councils of perfection. We know of no published examples that have

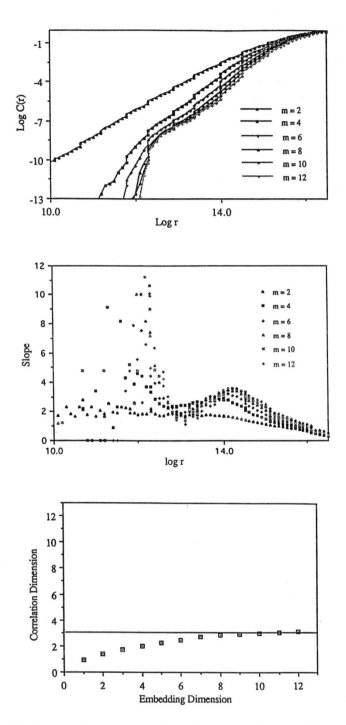

Figure 4.10: Determination of correlation dimension for the Great Salt Lake Data. (a) Plot of $C(r)$ again r; (b) Local slopes from (a); (c) Estimated dimension plotted against embedding dimension. From Sangoyomi et al. (in press).

followed such a complete program—but then,
very few published claims for "chaos" in natural time series have been accepted by the scientific community as well established. For com-
prehensive, overall reviews see Abarbanel et al.
(1993), Weigend and Gershenfeld (1994), Ott et
al. (1994), and Abarbanel (in press).

One problem with carrying out this program
is the availability of suitable software. A few
programs are provided in the disk accompanying these Notes, and sources for other programs
are listed in Appendix II.

Chapter 5

Forecasting Techniques, Underlying Physics, and Applications

David M. Rubin
U.S. Geological Survey
345 Middletfield Road MS 999
Menlo Park CA 94025 U.S.A.

5.1 INTRODUCTION

5.1.1 Preview

Science is based on the principle of repeatability: each time a system experiences similar conditions—both internal to the system and forces exerted externally on the system—we expect the system to exhibit a similar response. Forecasting exploits this principle by using the observed behavior of a system to predict behavior when similar conditions recur. Even if the equations describing a system are unknown, we can nevertheless use forecasting to learn about the system. For some purposes—such as weather forecasting, financial forecasting, or noise reduction—predicting the future is the primary goal of the forecasting. For the purpose of characterizing system dynamics, in contrast, predictions are made in an exploratory manner to learn what kinds of models perform best.

For a preview of how the forecasting procedure works, we can consider the Lorenz system described in Chapter 2. Three approaches could be used to predict the future of this system. First, we could measure the initial conditions (nonlinearity of vertical temperature gradient, temperature difference between rising and falling fluid, and intensity of convection) and use the three coupled equations (Equations 2.13) to predict the values of the three variables for successive steps in time.

A second approach could be employed if the governing equations were unknown, but sequential observations of the system were available. We could use the sequential observations to plot the 3-dimensional attractor (Figure 2.1), locate each predictee (a point whose three coordinates are given by the three variables that define the state of the system), identify nearby points (or neighbors) on the attractor, and predict the future state of the predictee by tracing the trajectories of the neighbors. For reasons that will be discussed below (see also Chapter 4) this same procedure can be employed on an attractor plotted using three lagged values of one variable rather than using simultaneous values of three variables.

The third approach requires even less knowledge about a system than the two outlined above. Even if we had no knowledge about the dimensionality of an attractor, we could perform exploratory computations for a variety of linear and nonlinear models and identify which models yield the most accurate forecasts. In this way we could not only predict the future of the system, but could characterize the system as well. Of course this procedure would have to be carried out computationally, rather than graphically, because there is no way to plot high-dimensional attractors in a two-dimensional image. This third approach forms the basis for the techniques that are discussed in detail below. Although these techniques rely on statistical operations, the techniques are rooted in several physical principles that are discussed below:

1. attractor trajectories can not cross (which would require that a system respond differently when identical conditions recur),

2. delay-coordinate embedding can be used to represent initial conditions, and

3. nonlinear relations can be approximated by local linear pieces.

5.1.2   Review and terminology of basic concepts

To those who are unfamiliar with the subject, it may seem contradictory that a system with sinusoidal behavior—and represented with a curved attractor—is purely linear (attractors and techniques for plotting them are discussed in Chapter 2). In this section we will begin by examining the math, physics, and attractor geometry of such a system. These concepts are important, because they provide the basis for understanding the nonlinear prediction methods that follow. An example of a linear system is a mass oscillating on a spring that exerts a compressional or extensional force that varies proportionally with the deformation of the spring. Such a system can be approximated by the equations

$$x_t = x_{t-1} + v_t \Delta t \qquad (5.1)$$
$$v_t = v_{t-1} + \frac{c_1 x_{t-1}}{m} \Delta t \qquad (5.2)$$

where x_t gives the location of the mass as a function of its previous location x_{t-1} and velocity v_t, Δt is a short step through time, c_1

is a coefficient that relates the force exerted by the spring to its displacement, and m is mass. Equation 5.1 defines the new location as the previous location plus the change in location during time Δt. Equation 5.2 defines the new velocity as the previous velocity plus the change in velocity during time Δt (rate of change in velocity, or acceleration, is equal to force $c_1 x_{t-1}$ divided by mass m). This system is described as linear because the future state of the system (location and velocity) is a linear function of the previous location and velocity. Although the system described by these equations is linear, the time series resulting from repeated iteration of the two equations is sinusoidal, and the attractor (plot of x_t vs v_t or x_t vs x_{t-1}) is curved (Figure 5.1). Linearity refers to the relation between variables, not to the shape of a plot of the time series or geometry of the attractor.

To define the state of this mass-spring system requires two variables: location and velocity. But two sequential values of location provide essentially the same information as simultaneous measurements of location and velocity, because velocity can be determined from the change in location through time; similarly, relative location can be determined from sequential velocities. Thus, the system state is uniquely defined by a combination of two variables (simultaneous location and velocity) or by sequential values of either variable. We can modify Equation 5.1 to describe the mass-spring system using two sequential values of location by substituting $(x_{t-1} - x_{t-2})/\Delta t$ for v_{t-1} in Equation 5.2 and then substituting the right side of Equation 5.2 for v_t in Equation 5.1:

$$x_t = x_{t-1} + \left[\frac{(x_{t-1} - x_{t-2})}{\Delta t} + \frac{c_1 x_{t-1}}{m} \Delta t \right] \Delta t \qquad (5.3)$$

(For more accurate iterative computational techniques, see Lorenz, 1963.) By rearranging terms and combining constants, Equation 5.3 can be simplified to

$$x_t = c_2 x_{t-1} + c_3 x_{t-2} \qquad (5.4)$$

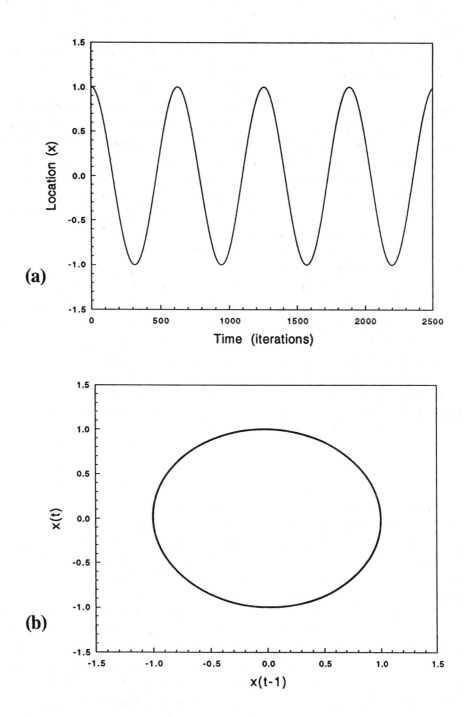

Figure 5.1: Linear mass-spring system described by Equation 5.1. (a) The time series is periodic and sinusoidal. (b) Attractor illustrated by plotting lagged values of location x.

which clearly illustrates the linear property of a sinusoidal time series: each value in the time series is a linear combination of two prior values.

One of the uses of forecasting is to evaluate the number of degrees of freedom of a system, which can be defined in at least two ways (Gershenfeld and Weigend, 1994). The *number of degrees of freedom* can be thought of as the number of variables that may be needed to uniquely define the state of the system mathematically. For example, the mass-spring system described by equation could be described as having two degrees of freedom (x and v in Equation 5.1; x_{t-1} and x_{t-2} in Equation 5.4). The equations have two other terms (c_1 and m in Equation 5.2; c_2 and c_3 in Equation 5.4), but, as long as the spring properties and mass do not change, these terms are constant and therefore do not contribute additional degrees of freedom. If, on the other hand, one or both of these terms vary periodically (as in an example discussed later in this chapter), then these terms would contribute additional degrees of freedom. One might argue that the *total number of degrees of freedom* of the mass-spring system is greater than 2, but that in the special case where the mass and spring are constant the *active number of degrees of freedom* is 2.

The distinction between active and total degrees of freedom is particularly important for geological systems. For example, the total number of degrees of freedom of a turbulent fluid is formally infinite (the continuous velocity, pressure, and temperature fields described by the Navier-Stokes equations). The active number of degrees of freedom, which is the number of degrees of freedom that can be evaluated using forecasting, may be considerably less than the total number. In the case of a fluid, for example, adjacent regions of the fluid do not necessarily behave independently. In such a case, the active number of degrees of freedom may be thought of as the velocity, pressure, and temperature for every small eddy having independent (uncorrelated) values from adjacent eddies. The purpose

in evaluating the number of degrees of freedom is to learn whether complicated behavior observed for a system arises because the system is complicated (many equations and many variables) or because of nonlinear properties of a simple system (few variables).

Equation 5.2 can be modified to describe a nonlinear system by modifying the term representing the force exerted by the spring. For example, force might vary with either the cube of the spring's displacement, requiring that $c_1 x_{t-1}$ be replaced with $c_1 (x_{t-1})^3$ (Figure 5.2). The resulting systems are nonlinear rather than linear, because x_t is no longer simply a linear function of previous values of x.

Both the linear and nonlinear mass-spring systems are periodic, but other linear and nonlinear systems may be nonperiodic. Nonperiodicity of linear systems is relatively easy to visualize; it originates where a system has so many degrees of freedom that initial conditions are not repeated; such a system is said to be stochastic. Nonperiodicity also can arise from a deterministic process: nonlinearity of the underlying equations can cause the system to exhibit behavior that evolves differently, even where initial conditions are nearly identical (Chapter 2).

It is easy to visualize how we might predict the future behavior of a simple periodic system (Figure 5.1) even in ignorance of the governing equations (Equations 5.1–5.4): we could merely use one wavelength of the time series as a template to predict the future, given any initial conditions. Two points are required to define the initial conditions, because all values (except the peak and trough) are encountered twice during each wave period—once during acceleration and once during deceleration; two sequential points distinguish between these two situations. Alternatively, we could make predictions by plotting the attractor in phase space and using the attractor to make predictions.

Adding a second periodic component to a time series (Figure 5.3) may make forecasting too complicated to be performed graphically,

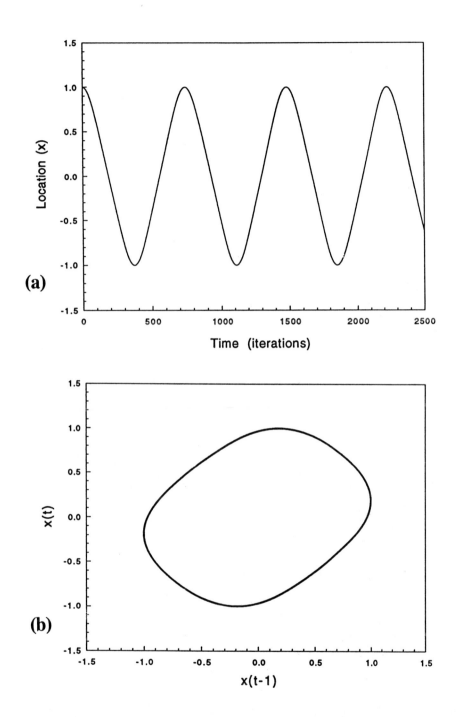

Figure 5.2: Nonlinear mass-spring system described by substituting $(x_{t-1})^3$ for x_{t-1} in Equation 5.2. (a) Time series of location x. (b) Attractor illustrated by plotting lagged values of x.

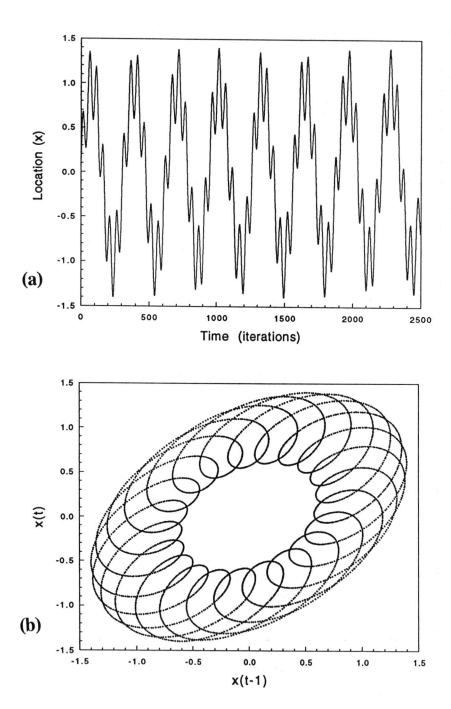

Figure 5.3: Plots of a two-frequency linear system. (a) Time series.(b) The attractor has the 3-dimensional shape of a torus; this image is the projection onto a plane.

particularly if the ratio of the two frequencies is not a rational number (quasiperiodicity). In such a case, the difficulty of making predictions is an indication of the limitation of the graphical predictive technique, not an indication that the system is particularly complicated. Such a system is simple (few degrees of freedom), linear, and retains a power spectrum that is diagnostic of periodicity (in this case, power is concentrated in two frequencies). The forecasting becomes more complicated for two reasons:

1. the relative phase of the two components may vary through time, causing the long-period cycles to differ from one to the next (either never repeating—as in the case of quasiperiodicity—or repeating after multiple cycles), and

2. the attractor dimension increases to three, making it difficult to represent clearly on a two-dimensional page.

The techniques presented in the following section are designed for forecasting in more complex systems with multiple frequency components (higher attractor dimension), noise, nonlinearity, and with external forces that vary through time.

5.2 TECHNIQUES

Several techniques have been developed to use the approach outlined above to extract information about the nonlinear structure of a time series or spatial image (Farmer and Sidorowich, 1988; Sugihara and May, 1990; Casdagli, 1992; Rubin, 1992; Theiler et al., 1992; Theiler et al., 1994; Casdagli and Weigend, 1994; Sugihara, 1994). The underlying principle of such time-series forecasting is to predict future values of a time series by consulting a catalog of how the system evolved at other times when initial conditions were similar. Predictions are made by selecting an event (predictee) with a known history, searching the catalog for one or more events for which the recent time-history approximates the time-history of the predictee, and then using the next values of these k nearest neighbors in the catalog to predict the next value of the predictee.

The idea of relating a sequence of past values of a time series to the future is based on physical principles, not merely statistical utility (Packard et al., 1980; Takens, 1981). The underlying principle is to use multiple values of a single variable to provide information about other variables that may be required to define the initial state of a physical system for which the future is to be predicted. As illustrated in the preceding discussion of Equations 5.1–5.4, sequential values of one variable (such as location) can provide information about other variables (such as velocity). For a rigorous proof of this concept of embedding, see Takens (1981).

For earth scientists who typically have a limited number of time series—often only a single time series—with which to understand a particular system, this ability to substitute one variable for several other variables seems almost too good to be true, if not pure magic. In practice, application of this principle is not so simple, because of the interaction of noise and delay time (time interval between values used in plotting an attractor, as defined in Chapter 4). If the chosen delay time is too short, the actual change in the time series is small, and even low levels of noise may mask the local structure of the attractor. On the other hand, if delay time is too large, then exponential divergence of trajectories means that the future state may have little or no relation to the first values in a sequence of lagged values used to represent the initial conditions.

Forecasting begins by splitting a time series into two pieces. One piece (a catalog or learning set) is used to relate the recent history of the series to the next value in the series. The other piece (testing set) is used to test the predictive ability of the learning set. In this forecasting process, the recent history of the system for m steps through time t can be represented by a single point in m-dimensional space; the coordinates of that point are $(x_{t-1}, x_{t-2}, x_{t-3} \ldots x_{t-m})$. To make each prediction, a predictee sequence of m values in the time series is placed (or embedded) in this m-dimensional space, and least squares is used to identify those m-dimensional sequences in the learning set that are closest to the predictee. This process is carried out computationally, but it can be visualized as plotting (in m-dimensional space) the point that represents the conditions for which a prediction is to be made and then locating the nearest points to this predictee; these nearby points represent other instances in the time series when conditions were most similar to the predictee. This process of locating these nearest neighbors can be visualized by another technique that is algebraically equivalent: graphically sliding the predictee sequence over a plot of the learning set time series and looking for the m-point sequences in the learning set that most closely match the predictee (for an example of this graphical approach, see Lendaris and Fraser, 1994). At least $m + 1$ of these nearest neighbors are located, so that least squares can be used to solve

$$x_t \approx \alpha_0 + \sum_{i=1}^{m} \alpha_i x_{t-1} \qquad (5.5)$$

for the $m + 1$ coefficients $(\alpha_0 \ldots \alpha_m)$ that best relate the future (x_t) to the past $(x_{t-1}, x_{t-2}, x_{t-3} \ldots x_{t-m})$ in the learning set. The second step in making each prediction requires that Equation 5.5 be solved again, this time substituting the coordinates of the predictee (a different sequence of $x_{t-1}, x_{t-2}, x_{t-3} \ldots x_{t-m}$) but retaining the same values for the coefficients $(\alpha_0 \ldots \alpha_m)$. This second solution of Equation 5.5 employs the relation determined from the learning set to predict x_t for the testing set. To predict each point in the testing set therefore requires that Equation 5.5 be solved twice (first to learn the values of the coefficients that best relate the past to the

future in the learning set, second to use those coefficients and the predictee sequence to forecast the next value in the testing set). Model performance is then evaluated by comparing the predicted values with the actual values in the testing set.

If the entire set of points in the learning set is used to evaluate the constants in Equation 5.5, then the technique is simply a multiple linear regression. In the nonlinear technique, a smaller number of (different) nearest neighbors in the learning set are used to re-evaluate the constants $(\alpha_0 \ldots \alpha_m)$ for each prediction, thereby effectively allowing Equation 5.5 to model nonlinear relationships using small locally linear pieces.

Much of the knowledge learned from forecasting is obtained by exploratory computations employing a variety of models with different embedding dimensions (number of sequential values used to quantify the initial conditions for each predictee) or different numbers of nearest neighbors (number of similar sequences in the learning set used to evaluate the constants in Equation 5.5 for each prediction). Two values can be used to quantify performance of these models:

1. correlation coefficient between predicted and observed values, and

2. the normalized RMS error.

E, the normalized RMS error of a model is given by

$$E = \frac{\left(\sum e_j^2 \right)^{1/2}}{\sigma} \qquad (5.6)$$

where j varies over the total number of predictions made for the testing set, e is the error for an individual prediction, and σ is the standard deviation of the time series; for the forecasting technique to be meaningful, the time series must be stationary, requiring that the standard deviation be equal for the testing and learning sets.

A model that predicts each value perfectly has a normalized RMS error of zero and a correlation coefficient of 1.0, whereas a model that repeatedly predicts the mean value of the time series has a normalized RMS error of 1.0 and a correlation coefficient of zero. In practice, however, the correspondence between RMS error and correlation coefficient is not unique. For example, if each predicted value is exactly double the observed value, the correlation coefficient is 1.0, but the normalized RMS error is greater than zero. For this reason, RMS error is a more precise estimate of model performance. The correlation coefficient is nevertheless a useful measure of model performance, particularly where it is desirable to quantify the ability of the model to explain the variance of the time series (the amount of variance explained by the model is equal to the square of the correlation coefficient).

One application of exploratory modeling is Casdagli's (1992) deterministic-versus-stochastic forecasting technique, which measures forecasting error as a function of the number of neighbors (similar events) used to make predictions. At one extreme (stochastic linear modeling), forecasts are based on behavior learned from all events in a learning set. This kind of global linear regression model maximizes noise reduction but minimizes sensitivity to the specific initial conditions for the event that is being forecast. At the other extreme (deterministic nonlinear modeling), forecasts are based on the relations learned from a small number of events for which the initial conditions are most similar to the event that is being forecast. In these nonlinear models, noise reduction is poorer, but sensitivity to initial conditions is enhanced. Casdagli (1992) argued that the dynamics of a system can be characterized by the class of model that makes the most accurate short-term forecasts. Low-dimensional nonlinear nonperiodicity (chaos) can be identified in systems where nonlinear models employing a small number of nearest neighbors and

small embedding dimension outperform global linear models.

In an exploratory search for nonlinearity, it may be advantageous to vary the forecasting time (the number of time steps from the end of the predictee sequence to the time for which a value is predicted), because the relative improvement of a nonlinear model may become more evident at prediction times greater than a single time step. As the prediction time continues to increase, all models may perform so poorly that it becomes difficult to detect an advantage of any model. In Equation 5.5, the subscripts indicate a single step-ahead forecast, but the equation can be modified to describe a forecasting time of n time steps by substituting $x_{t+1-(i+n)}$ for x_{t-i}.

Another approach in forecasting has been to compare forecasts of an original time series with forecasts made from surrogate series (Theiler et al., 1992). The surrogates are created to mimic some, but not all, attributes of the original. For example, surrogates made to have power spectral magnitudes as the original—but having randomized phases—can be used to test the null hypothesis that the original time series is linearly correlated noise. If the original and surrogate time series have significantly different forecastability, then this hypothesis can be rejected; such results demonstrate that the original time series has a deterministic nonlinear structure that is not retained in the surrogates.

Similarly, forecasting accuracy can be measured for models that vary the embedding dimension m, to evaluate the number of active degrees of freedom of a system from which a time series was sampled. For example, m must be at least three to accurately forecast the behavior of a system with three degrees of freedom, such as Lorenz's (1963) simplified model of convection (Equation 2.13). If the value of m used to make forecasts for this system is less than three, then trajectories cross (as in the two-dimensional image of the attractor of the Lorenz system in Figure 2.1). Consequently, forecasts made for an intersection may be inaccurate, because the wrong trajectory is chosen. Increasing the embedding dimension eliminates these intersecting trajectories, and is said to "disambiguate" the data; accurate forecasts can now be made. By applying forecasting in an exploratory manner (performing computations to evaluate the relative performance of models with different embedding dimensions), a lower limit of the number of degrees of freedom can be evaluated. In this application, forecasting performs the same function as the false neighbors technique described in Chapter 4.3.

Sugihara and May (1990) suggested using the decay (with time into the future) of forecasting accuracy to distinguish uncorrelated noise from chaos in time series. The idea behind this technique is that chaos can be predicted for short times into the future, whereas uncorrelated noise is unpredictable even for a single step in time. Although this technique can recognize the null case of uncorrelated noise, some of the more interesting null hypotheses (correlated noise or combinations of deterministic structure and correlated noise described by Rubin, 1992) can not be distinguished from chaos using this technique.

Additional details of these modeling techniques are given by Casdagli (1992) and Casdagli and Weigend (1994); a computational algorithm for spatial forecasting is described by Rubin (1992). The knowledge to be gained from these forecasting techniques can be compared to that gained from spectral analysis. Both techniques provide information about how a system operates, but neither provides the specific equations that describe the system. Determining that a particular nonperiodic system is linear or nonlinear, like determining the dominant frequencies of a periodic system, is merely one step in characterizing or understanding the system.

5.3 COMPLICATION OF UNSTEADY FORCING

5.3.1 Nature of the complication

The preceding techniques are based on the assumption that forcing is constant (external forces applied to a system are constant through time). Where this condition is met, the observed history of the system results purely from intrinsic processes or self-organization. Although this condition can be approximated in lab experiments, few geologic systems meet this requirement. If forecasting techniques are applied to a system subject to unsteady forcing, the results may apply to the external forces rather than to the system that is being studied. For example, if the stiffness of the spring in our mass-spring system (c_1 in Equation 5.2) varied in response to oscillations in temperature, the system would be altered in two important aspects:

1. the observed history of the system would be grossly different (Figure 5.4), and

2. some of the observed history would reflect the character of the forcing rather than the character the mass-spring system.

We can use the same approach to modify the Lorenz equations (Equations 2.13) to describe convection resulting from unsteady forcing (time-varying temperature differential between the upper and lower surfaces of the fluid). A time-varying temperature differential is incorporated by allowing r (the ratio of the Rayleigh number to the critical value for the initiation of convection) to vary from one time step to the next. As in the mass-spring system, this unsteady forcing results in a time series that reflects both system behavior and forcing (Figures 5.5 and 5.6).

Complications due to unsteady forcing are ubiquitous in geologic systems. For example, a time series of sediment transport at a point on a bed of ripples in a tidal flow would have

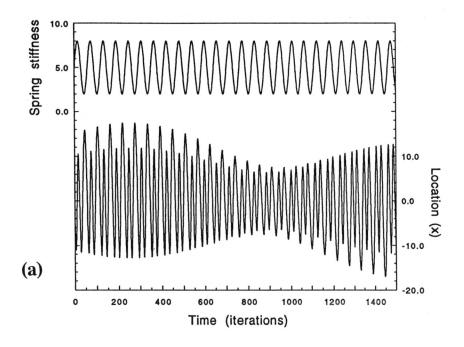

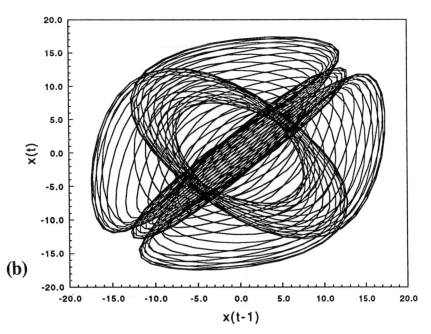

Figure 5.4: Unsteady, linear, mass-spring system described by allowing spring stiffness (c_1 in Equation 5.2) to vary sinusoidally through time. (a) Time series of spring stiffness and location x. (b) Attractor.

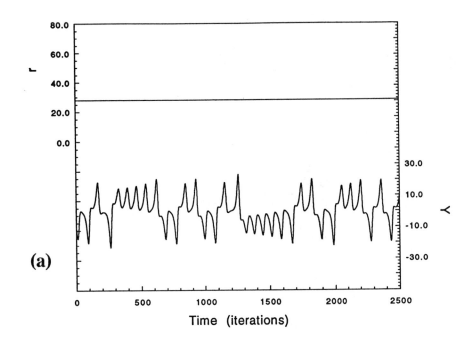

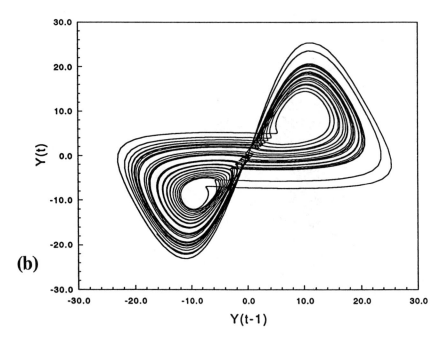

Figure 5.5: Steady Lorenz system. (a) Time series of Y (temperature difference between ascending and descending fluid) computed using the standard value of $r = 28.0$ in Equation 2.13. (b) Attractor of system in (a).

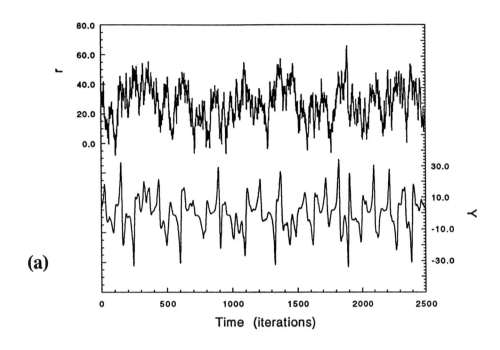

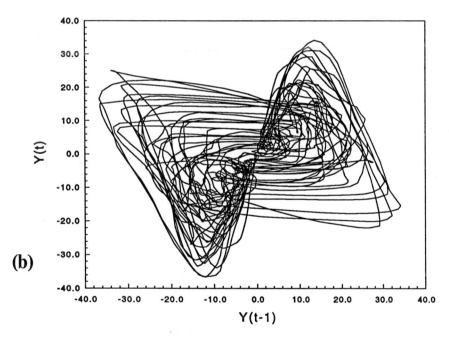

Figure 5.6: Unsteady Lorenz system. (a) Time series of Y incorporating a time-varying r. The mean value of r is 28.6, approximately the same as in Figure 5.5a. The time series of Y is more complex because of the unsteady forcing (r) applied to the system. (b) Attractor of system in (a).

two components: the cyclicity of astronomically induced tidal flow as well as transport variations caused by the passage of the ripples on the bed. The ripples would display a tidally driven cyclic behavior in addition to any self-organized interactions between ripples. If the processes operating in this system were completely unknown (and the periodic forcing was therefore not recognized) an investigator might be misled into thinking that ripples have an intrinsic tidal cyclicity, whereas the tidal cyclicity merely reflects the forcing.

The importance of external forcing in the system described above is so obvious that it may seem absurd to worry about overlooking it. But the effects of unsteady forcing may be much more obscure in geologic data. For example, the structure of a stratigraphic sequence might be influenced by:

1. processes within the depositional environment (analogous to behavior of the Lorenz system under steady forcing),

2. changes in conditions in adjacent regions (analogous to changing the size or shape of the fluid body), or

3. global changes such as climate (analogous to a universal change in the temperature difference between bottom and top of the fluid).

In modern systems, three approaches can be used to work around the problem of unsteady forcing:

1. regulate forcing experimentally,

2. measure forcing as well as system response, and only use data collected at times when forcing is within a narrow range, and

3. measure forcing as well as the system response and use the measured forcing as input in the modeling (input-output modeling of Hunter and Theiler, 1992).

In the following examples, the complication of unsteady forcing was resolved by keeping forcing constant (Lorenz example), by choosing a field site where forcing was spatially uniform (wind-ripple example), and by incorporating the unsteady forcing in input-output models (climate and surf-zone sediment transport). In geologic time series, however, the problem of unsteady forcing may be intractable, because only response (not forcing) can be inferred from stratigraphic deposits.

5.3.2 Input-output modeling

Input-output modeling (Hunter and Theiler, 1992) is particularly useful in the earth sciences, where it is usually impossible to regulate the external forces exerted on a system. Instead, input-output modeling utilizes two simultaneous time series: forcing (input) and system response (output). The underlying principle is to use a catalog to learn how the system responds to different forcing events. The technique is computationally similar to the single-series forecasting described in Equation 5.5, but it relates response x at time t, to the forcing y measured during a sequence of m steps through time:

$$x_t \approx \alpha_0 + \sum_{i=1}^{m} \alpha_i y_{t-i} \qquad (5.7)$$

The response of a system may lag behind the forcing, and forecasting can be used to quantify such a lag. Equation 5.7 can be modified to incorporate such a lag by replacing $y_{(t-i)}$ with $y_{(t-(i+n))}$, where n equals the lag time from the end of the sequence of input forcing values to the time of the model-response output. Equation 5.7 is applied in the same manner as Equation 5.5: once to solve for the local linear relation between forcing and response in a learning set, and a second time to predict the response for the testing set. The lag of a physical system is quantified by determining the value of lag n that yields the most accurate forecasts. An example of this approach is given in the discussion of surf-zone sediment transport (see below, Subsection 5.4.5).

5.3.3 Forecasting in practice

In applying the forecasting techniques, a researcher might go through the following sequence of operations to characterize the system that produced an observed time series (Table 5.1):

1. Use spectral analysis to evaluate the time series for periodic or nonperiodic structure. If the power of the time series is restricted to one or more narrow frequency bands, the time series is largely periodic and linear. Although the time series may still contain a minor nonlinear component, the nonlinear techniques will probably not greatly improve the forecasts. In contrast, if the power of the time series is distributed across broad bands of frequencies, then nonlinear techniques may discover a deterministic nonlinear structure hidden in what the spectral analysis detects as "noise." Although spectral analysis is a useful technique for identifying periodic (linear) structure in a time series, this technique can not identify structure that is nonperiodic or nonlinear. Nonlinear techniques are useful where a time series contains nonlinear structure.

2. Split the time series into a testing set and learning set. To reduce computation time, the testing set should have just enough points to obtain stable error statistics (values that do not change appreciably if the number of points were to be increased); a few hundred points is generally sufficient, provided that the testing set and learning set both are representative samples of the underlying attractor.

3. Choose the first exploratory values for embedding dimension (m), number of neighbors (k), and forecasting time (n), and apply Equation 5.5 (5.7 for input-output systems) to every point in the testing set. Use Equation 5.6 or the correlation coefficient between predicted and observed values to evaluate the performance of that model. Repeat this procedure for other values of m and k, searching for the model that makes the most accurate predictions.

4. To demonstrate nonlinearity, the best nonlinear model should be substantially more accurate than the best linear model (for a given forecasting time). An additional test for nonlinearity is to use surrogate data, as explained above, and as will be illustrated in the wind-ripple example below.

5. The success of modeling efforts depends on how well null hypotheses are formulated and eliminated. In particular, low-dimensional nonlinearity (chaos) can only be demonstrated by formulating and eliminating null hypotheses that include interactions between noise and periodicity as well as the more common null hypotheses of high-dimensional linear and nonlinear systems.

5.4 EXAMPLES

5.4.1 Lorenz system

Before applying forecasting techniques to geophysical data, it is instructive to examine results for mathematical models such as the Lorenz system (Chapter 2) for which the underlying dynamics are known. Using a time series of a single variable (such as the temperature difference term), we can use forecasting to demonstrate that the system is nonlinear and deterministic, with 3–4 degrees of freedom.

We begin with a time series of the temperature-difference variable computed from the Lorenz equations. Spectral analysis of this time series demonstrates that it is nonperiodic

Table 5.1: Forecasting techniques and their application.

INTERPRETIVE GOAL	TECHNIQUE	REFERENCES
Estimating degrees of freedom (lower limit)	Measuring forecasting error as a function of embedding dimension(number of successive past values used to predict the future state)	Weigend and Gershenfeld (1994)
Distinguishing stochastic and deterministic structure	Measuring forecasting error as a function of number of nearest neighbors	Casdagli (1992) Casdagli and Weigend (1994)
Identifying nonlinear structure	Comparison of forecasting errors of original and surrogate data

Measuring forecasting error as a function of number of nearest neighbors | Theiler et al. (1992)

Casdagli (1992) Casdagli and Weigend (1994) |

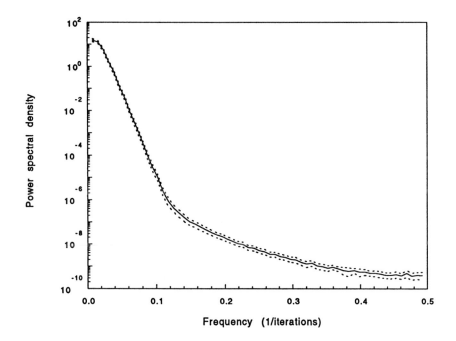

Figure 5.7: Power spectral density for a 6000-point time series of the Lorenz system; nonperiodicity is demonstrated by the lack of a peak. The solid line represents power spectral density, and the dotted lines represent the 95% confidence limits (calculated from the standard deviation of power at each frequency). To perform these calculations the time series was broken into 128-point pieces.

(Figure 5.7), which makes the time series a candidate for nonlinear forecasting. To perform the forecasting, approximately 3000 points are used for a learning set, and several hundred points are reserved to test the predictive ability of the relations learned in the learning set. We then evaluate model performance as a function of embedding dimension (the number of preceding points that are used to predict each next value in the time series), prediction time, and the number of nearest neighbors used to evaluate the coefficients in Equation 5.5.

Results of exploratory modeling demonstrate that model performance is optimum if the embedding dimension has a value of 4 (Figure 5.8). This value is larger than the actual number (3) of degrees of freedom of the Lorenz system, because modeling error and noise (even digitization error!) can preferentially degrade models using low embedding dimensions. Consequently, the embedding dimension of the most accurate model is an upper limit of a system's number of active degrees of freedom. Any stationary nonperiodic system must have at least 3 degrees of freedom; if a stationary system has only two dimensions, then it can not be nonperiodic because its attractor must be confined to a single plane. Trajectories can not cross each other, because this would imply that a system behaves differently for exactly the same conditions; the trajectories therefore must either be attracted to a fixed point, spiral outward indefinitely (non-stationary), or follow a single repeating (periodic) path. The Lorenz system does not follow any of these paths, so we conclude that the system has at least 3 active degrees of freedom. Thus, the number of degrees of freedom must be greater than or equal to 3 and less than or equal to 4.

Results also demonstrate that the Lorenz system is most accurately modeled using a small number of nearest neighbors (Figure 5.9), which indicates that the system is situated toward the deterministic end of the deterministic-stochastic spectrum. The best model is one that learns from the behavior of only the nearest 16 neighbors to each predictee; the best model therefore ignores more than 99% of the learning set when making each prediction.

Because real data are not as noise-free as computer-generated data, it is instructive to add noise to the Lorenz time series and repeat the forecasting computations. For this experiment we added uncorrelated noise having a standard deviation of 15% of that of the signal (the Lorenz time series). Noise alters the results in the following manner: forecasts are poorer, and—more significantly for characterizing the system—the best model employs a greater number of nearest neighbors (because the Lorenz-plus-noise system is more stochastic) and a larger embedding dimension (because of the noise-reducing capability of the longer sequence of previous points). These results emphasize two of the points made above. First, the embedding dimension of the best model is merely an upper limit to the number of degrees of freedom in the system. Second, a natural time series is influenced by multiple processes:

1. behavior of the system of interest,

2. the effects of unsteady natural forcing exerted on the system of interest, and

3. artificial processes such as measurement error and instrument noise.

Forecasting may not be able to distinguish these individual components.

We might attempt to use forecasting to document the exponential divergence of nearby trajectories in the Lorenz attractor, but simple attempts would be unsuccessful. Individual trajectories stretch apart and are compressed with each orbit around the attractor, making it impossible to document a systematic exponential divergence from only a few trajectories. Alternatively, the error statistics computed from Equation 5.6 are averaged in such a way as to obscure exponential divergence. These computational difficulties are in addition to the geome-

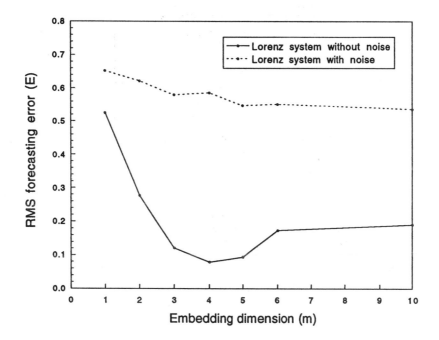

Figure 5.8: RMS forecasting error as a function of embedding dimension for the Lorenz system with and without 15% white (uncorrelated) noise. The noise-free time series is modeled most accurately using an embedding dimension of four, but the system actually has only three degrees of freedom. Forecasts for the noisy series are less accurate, and the series is modeled most accurately using a higher embedding dimension.

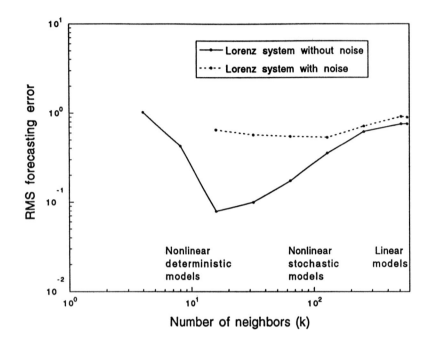

Figure 5.9: Forecasting error as a function of number of neighbors (k) used to solve Equation 5.5 for the Lorenz system with and without 15% white noise. Where k is small, forecasts are nonlinear and deterministic; where k is large, models are linear and/or stochastic (Casdagli, 1992). The relative performance of these deterministic-versus-stochastic models can be used to infer properties of the system dynamics. Where minimum prediction errors are for models with low k, the system is deterministic and nonlinear. The noise-free series is modeled most accurately using models near the nonlinear deterministic extreme, whereas the noisy series is modeled more accurately using nonlinear stochastic models.

try of the attractor, which limits divergence by confining the trajectories to a small region of space.

5.4.2 Turbulent lateral separation eddies

Lateral separation eddies are fixed eddies that occur in channel expansions (Figure 5.10). The main channel flow separates from the bank at the upstream end of the eddy and reattaches to the bank at the downstream end. Where the flow reattaches, water piles up, and the flow within the eddy is driven upstream. Field and lab measurements demonstrate that these eddies pulsate erratically even when subject to constant forcing (constant main-channel discharge); spectral analysis demonstrates that the pulsations are nonperiodic (Rubin and McDonald, 1995).

At least three classes of systems can exhibit nonperiodic behavior:

1. high-dimensional (many degrees of freedom) linear systems,

2. low-dimensional nonlinear systems (chaos), and

3. high-dimensional nonlinear systems.

Ideally, we would like to characterize nonperiodic eddy pulsations as one of these kinds of systems. In the first half of this century, Landau (1944) proposed that turbulence (nonperiodic flow) resulted from a large number of modes of excitation of a fluid, but in the last few decades, it has been shown theoretically and experimentally that some examples of nonperiodic flow are low-dimensional chaos (Lorenz, 1963; Ruelle and Takens, 1971; Gollub and Swinney, 1975; Gollub et al., 1980). The experimental studies that have documented low-dimensional chaos have focused on flows that are at the threshold of turbulence (transitional with laminar flow).

Several alternate explanations that do not involve low-dimensional chaos (spin-glass relax-

ation, spatial noise amplification, and transients) have been proposed recently and are noted by Crutchfield and Kaneko (1988). They argue that transient effects can dominate a system for long time intervals, and they therefore question the relevance of low-dimensional chaos to fully developed turbulence.

Low-dimensional chaos (nonperiodicity with few degrees of freedom) such as the Lorenz system requires both low-dimensionality (by definition) and nonlinearity (to allow nonperiodicity in a low-dimensional system). Two forecasting techniques can be used to evaluate the hypothesis that nonperiodicity in recirculating eddies results from low-dimensional chaos. More precisely, the techniques are an attempt to falsify the null hypothesis that the observed flows are high-dimensional systems as proposed by Landau, 1944. First, we can use the deterministic-versus-stochastic technique to look for nonlinear structure in the time series. The best models (those with the lowest error) for the data set are purely linear (Figure 5.11), which does not contradict the hypothesis that this turbulence results from linear stochastic processes.

The second technique is an attempt to determine the number of degrees of freedom of the lateral separation eddy. The best forecasting models employ an embedding dimension approaching 100, too high to support the idea of low-dimensional chaos (and probably too high to be meaningful for the length of the time series). Thus, this result fails to contradict Landau's hypothesis that turbulence is high-dimensional, and the deterministic-versus-stochastic technique fails to identify nonlinearity.

In some situations, low-dimensional chaos can be masked by stochastic effects such as noise or measurement error. Stochastic effects resulting from as little as 10% measurement error can be sufficient to mask low-dimensional dynamics (Rubin, 1992). In the recirculating flows in the lab, the flow forcing and measurement error had a standard deviation of only 2% of the

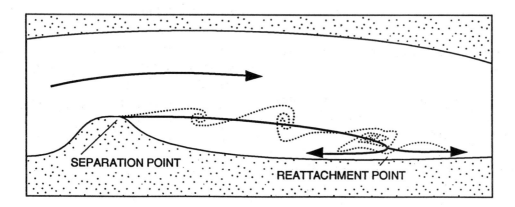

Figure 5.10: Schematic diagram of a lateral separation eddy.

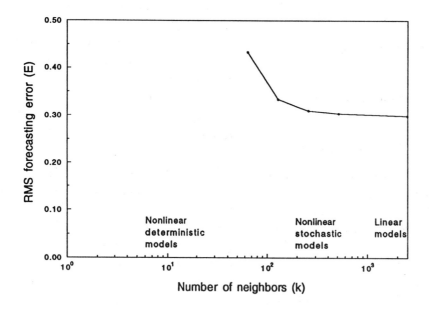

Figure 5.11: Forecasting error as a function of number of neighbors for a time series from a pulsating lateral separation eddy in the Colorado River, Grand Canyon (Rubin and McDonald, 1995). This time series is modeled most accurately using linear stochastic models. The time series was recorded during a time when release from Glen Canyon Dam (approximately 220 km upstream) was nominally steady, and stage-gauge data indicate that discharge 96 km upstream from the measurement site was 430 m^3/s with a standard deviation of 2 m^3/s.

mean, suggesting that the nonperiodicity results from high-dimensional flow processes (linear or nonlinear) rather than from unsteady forcing or measurement noise.

5.4.3 Dripping handrail spatial patterns

Forecasting can be applied to spatial patterns as well as to time series (Rubin, 1992). For this purpose, Equation 5.7 can be modified to:

$$z_{(x,y)} \approx \alpha_{(0,0)} + \sum_{i=0}^{h-1} \sum_{j=0}^{w-1} \alpha_{(i+1,j+1)} z_{(X+i,Y+j)}$$

$$(5.8)$$

where $z_{(x,y)}$ is the value of the pixel to be predicted, x and y are coordinates measured relative to the lower left corner of the image, h and w are the height and width of a plaquette (rectangular region) of pixels used to make each prediction, and X and Y are the coordinates of the lower left pixel in the plaquette (Figure 5.12). For the examples presented here, each predicted pixel is centered left-to-right with respect to the plaquette ($X = x - (w-1)/2$). The distance from the pixel that is being predicted to the closest pixel in the plaquette can be equal to 1 pixel ($Y = y - h$, as in Figure 5.12 and the computations for Figure 5.13) or can be greater (as in the case of the wind-ripple computations discussed later). In a spatial pattern, the embedding dimension m is equal to wh (the number of pixels in each plaquette); the subscripts of $\alpha, (0,0) \ldots (w,h)$, identify the $wh+1$ regression coefficients.

By applying equation 5.8 to an image of the computer-generated dripping handrail image (Figure 2.8), we can extract information about the processes used to create the image. For example, for a prediction distance of 1 pixel, forecasting error is a minimum for a plaquette having a height of 1 and a width of 3 (Figure 5.13), precisely the dimensions of the algorithm used to create the image. For longer prediction distances, the best predictions are made using a larger plaquette, to account for water that flows along the handrail (in or out of the left and right sides of the plaquette).

The nonlinearity of the handrail algorithm can also be documented using exploratory forecasting: deterministic nonlinear models outperform the linear models by approximately 5 orders of magnitude (Figure 5.14). If uncorrelated noise is added to the dripping handrail image, forecasting accuracy decreases, particularly for nonlinear forecasts. Noise levels approaching or exceeding 10% preferentially degrade the nonlinear predictions to such an extent that nonlinear models become less accurate than linear models (Figure 5.14). Evidently, sensitivity to initial conditions of the nonlinear model has become less an advantage than noise-reducing capability of the linear model.

5.4.4 Wind-ripple geometry

Most studies of ripple and dune geometry have been directed at determining how mean geometry varies for differing flow conditions, but another question is equally interesting: what causes differences in geometry of sequential bedforms in a train created by a uniform, steady flow. Hypotheses to explain this variation include quasiperiodicity, randomness, or deterministic chaos resulting from modification of the flow or sand-transport field by the bedform immediately upstream (Rubin, 1992).

Most previous models of ripples and dunes have treated such bedforms as periodic (Allen, 1978; Rubin, 1987) or random (Paola and Borgman, 1991), but several morphologic and behavioral characteristics suggest that the complexity is self-organized. Even where flow is uniform (when averaged on a scale that is large relative to individual ripples), geometric variation of bedforms is ubiquitous. A deterministic cause for this complexity is suggested by experiments and computational models demonstrating that ripples perturb the local boundary layer or sand-transport field, thereby modifying

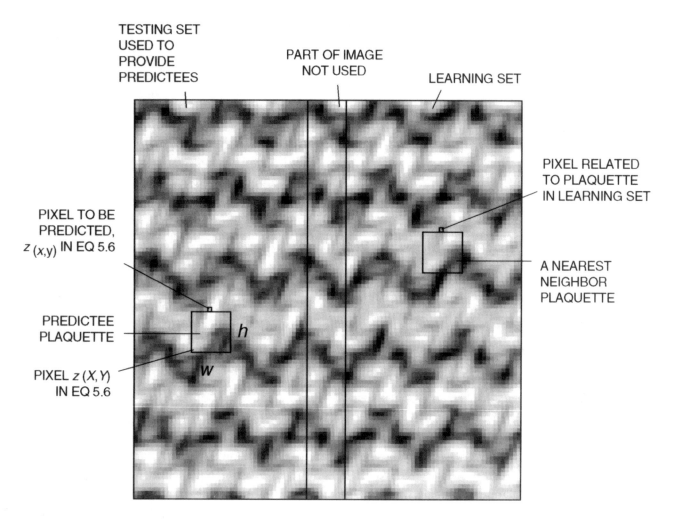

Figure 5.12: Diagrammatic representation of the spatial forecasting procedure. An image is divided into two parts, one part for use as a learning set and one half to test predictions. A small area between these two regions is not used in either set, to avoid lateral correlation. The image in this example is 100 pixels square; predictee plaquettes are 10 pixels square ($h = 10, w = 10$), corresponding to an embedding dimension of 100. Each predictee plaquette is compared with all 10-by-10 plaquettes in the learning set, and nearest neighbors are evaluated by the least squared differences of corresponding pixels in the predictee and learning set plaquettes. Predictions are made using Equation 5.8. The predictive process is illustrated using a two-dimensional image, but the procedure is computationally identical to representing each plaquette as a point in 100-dimensional space, finding nearest neighbors in this space, and making predictions by tracking the trajectories of those points to their future locations. The image being forecast is a gray-scale image of a quasiperiodic pattern created by superimposing 3 trains of sine waves having differing orientations, wavelengths, and planform sinuosities; values of pixels range from zero (white) to 255 (black).

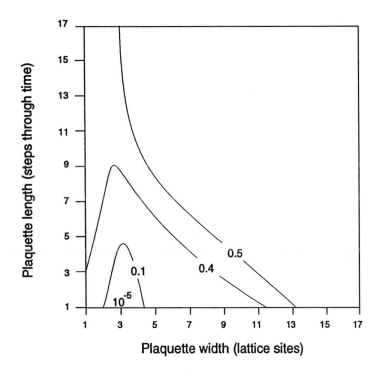

Figure 5.13: Forecasting error as a function of embedding-dimension height h and width w for the dripping handrail image (Figure 2.8). One-step-ahead forecasts for this spatial pattern are most accurate using plaquettes 3 pixels wide and 1 pixel high, corresponding to the known dynamics of the equations used to create the image.

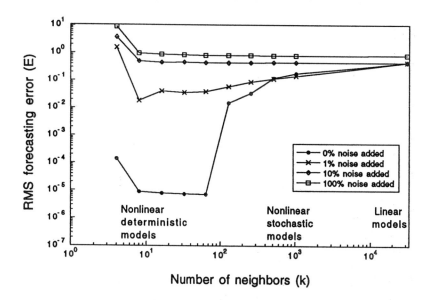

Figure 5.14: Forecasting error for the dripping handrail, with varied amounts of additive uncorrelated noise (from Rubin, 1992). Forecasts (208 predictees) are for distances of 1 pixel and are based on plaquettes 3 pixels wide by 1 pixel high. For noise levels of 0% and 1%, the best models are nonlinear, employing few of the 19,656 neighbors in the learning set. For noise levels of 10% and 100%, errors are much greater, and the best models are at the linear stochastic extreme.

the conditions that shape the next ripple downstream (Raudkivi, 1963; Southard and Dingler, 1971; Anderson, 1990; Forrest and Haff, 1992). In wind, ballistic grain impacts are believed to be more important than fluid effects, and self-organization has been suggested to arise from a sorting process that causes adjacent ripples to attain similar sizes and migration speeds (Anderson, 1990; Forrest and Haff, 1992).

In flowing water, downstream coupling occurs because the local flow near the bed is influenced directly by the bedform immediately upstream. In some flows, placing a single artificial ripple or obstacle on a flat bed can be sufficient to induce formation of a train of ripples downstream (Raudkivi, 1963; Southard and Dingler, 1971), even in a flow where ripples otherwise would not form (Figure 5.15 a). Similar spatial patterns can be simulated using the dripping handrail model starting from initial conditions that are uniform except for a slight perturbation (Figure 5.15 b). In the model, each horizontal row in the image represents a step through time and is computed from the conditions at the previous time. In the real ripple pattern, each row represents an increasing distance downstream and develops in response to upstream conditions. In both the real and computational examples, the patterns that develop illustrate a similar sensitivity to initial (previous or upstream) conditions.

A complete treatment of ripple dynamics requires an additional dimension of complexity beyond that implied above. The discussion above implies that the downcurrent ripple geometry is a function of upcurrent geometry-unchanging through time. Although this is true for some flows (as illustrated in Figure 5.15 a), in most flows ripple dynamics is more properly considered as a two-dimensional spatial system that evolves through time as individual ripples interact while migrating downcurrent.

In some ripple fields, down-current coupling produces current-parallel lanes of ripples with abrupt discontinuities between lanes (Allen,

1968). Such structure implies that across-current coupling of the real ripples is weak relative to down-current coupling, an hypothesis that is supported by computational experiments with the dripping handrail model. In addition to these reported spatial variations in individual ripples in uniform flows, both field observations and laboratory experiments have found that the planform geometry of dunes becomes more complex as flow strength increases (Allen, 1968; Middleton and Southard, 1984). This increasing spatial complexity is analogous to the increasing complexity observed at increasing flow strengths in Couette cylinders (Gollub and Swinney, 1975) and convection cells (Gollub, Benson, and Steinman, 1980).

A photograph of ripples formed by wind (Figure 5.16) was digitized and analyzed using the spatial forecasting techniques discussed above; pixel intensity was used to represent ripple geometry. Unlike the synthetic images, in which intensity is proportional to elevation, intensity in the digitized photograph image is more nearly a function of slope.

As in the case of one-dimensional time series, forecasting errors of spatial patterns vary with embedding dimension (Figure 5.17). The best forecasts are obtained using a plaquette with a height of 9 pixels (slightly greater than the mean wavelength of 8.3 pixels) and a width of 5 pixels. This result is in good agreement with the hypothesis that the geometry of any one ripple depends on the geometry of the ripple immediately upstream.

Forecasting error was measured as a function of the number of nearest neighbors used to make the predictions (Figure 5.18). The ripple pattern is forecast most accurately using a nonlinear stochastic model, which offers a 60% improvement over the best linear model. This property of the forecasting errors suggests that the ripple pattern contains a nonlinear component.

Forecasting errors for the original image differ significantly from surrogate images that were

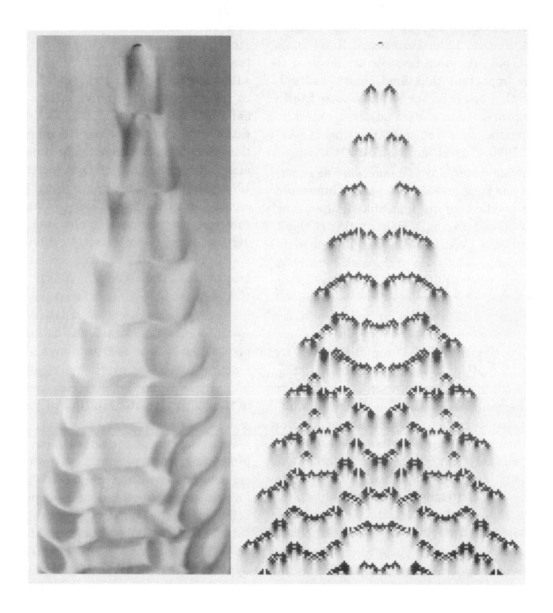

Figure 5.15: Development of instabilities beginning from a slight artificial perturbation (from Rubin, 1992). (a) Planform view of an experiment to create ripples on a flat bed in flowing water. No sand-transport occurred until a mound of sand was placed on the bed. The mound disturbed the flow, producing another bedform, which in turn disturbed the flow, and so on downcurrent (from top to bottom). Photograph from Southard and Dingler (1971), digitally processed to remove perspective distortion. (b) Computer simulation of spatial differences using the dripping handrail model, beginning with initial conditions (top of image) that were uniform except for a single pixel. Image is a re-computation of Crutchfield and Kaneko (1987, fig. 45). In both (a) and (b), conditions at the top of the image determine the structure at lower locations.

Figure 5.16: Photograph of ripples formed by wind blowing over a sand bed that is otherwise flat (Rubin, 1992). Wind direction is from top to bottom; ripple wavelength is approximately 10 cm (corresponding to 8.3 pixels after the photograph was digitized). The digital version of this image was used as input for the forecasts in Figures 5.17 and 5.18.

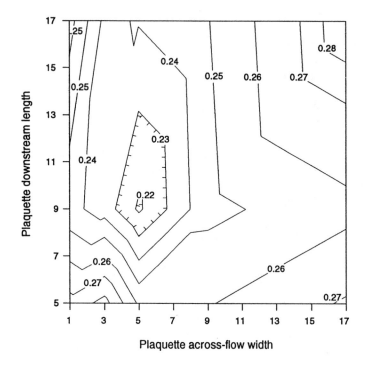

Figure 5.17: Predictability of wind ripples in Figure 5.16 as a function of embedding dimension (from Rubin, 1992). The best forecasts are based on plaquettes that extend approximately 1 wavelength downwind and one-half wavelength across-wind.

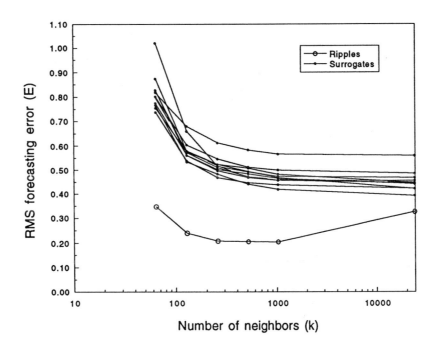

Figure 5.18: Forecasting error as a function of number of nearest neighbors for wind ripples and 11 surrogates (from Rubin, 1992). Plaquette size is 9 pixels (downwind) by 5 pixels (across-wind), as determined to be optimum in Figure 5.17. The best nonlinear model offers a 60% improvement over the best linear model, and the original differs considerably from the surrogates (values of sigma as large as 8).

created to have the same FFT magnitude but randomized phases. The difference is as great as 8 sigmas (Theiler et al., 1992). That is, the difference between the error for the original and the mean error for the surrogates is 8 times the standard deviation of the errors for the surrogates (Figure 5.18). These results (most accurate predictions with a nonlinear model and significant differences in forecasting error between the real and a linear surrogate pattern) are inconsistent with the null hypothesis that the structure is linear and non-deterministic.

5.4.5 Surf-zone transport

Despite considerable study, virtually no sediment transport models are capable of accurately predicting the instantaneous suspended sediment concentration in response to wave-generated currents in the surf zone. Most models of unsteady sediment transport merely treat the unsteady flow as a time-varying river, and use relations developed for steady flow to predict sediment concentration; any such approach must result in a predicted concentration time series that mimics the associated velocity time series. Merely inspecting simultaneous time series of near-bed flow velocity and sediment concentration demonstrates that this is not the case (Figure 5.19); concentration shows no visible relation to the flow. As might be expected for time series with such an appearance, the correlation coefficient between sediment concentration and flow velocity (or absolute value of velocity) is zero. One hypothesis to explain this lack of predictability is that the system is noisy. For example, sediment might become suspended somewhere upstream and then be advected to the measurement device. Although such unpredictable processes probably occur, the input-output techniques described above can be used to predict sediment concentration relatively accurately, which demonstrates that concentration is controlled by the flow but that the relation is more complicated than previously believed (Jaffe and Rubin, 1995).

To make accurate predictions requires using three techniques described above:

1. embedding (relating concentration to a sequence of velocity measurements rather than to a single value),

2. nonlinear deterministic modeling (using a nonlinear relationship—a small number of nearest neighnors—to relate forcing to response in Equation 5.7), and

3. incorporating a lag time between the input (flow) and output (sediment concentration) measurements.

This approach leads to models that can predict concentration relatively accurately—the correlation coefficient between predicted and observed concentrations is 0.6. The models even predict pulses of suspended sediment at times when velocities are relatively weak, as illustrated in Figure 5.19.

In retrospect, the benefits of all three modeling essentials can be appreciated intuitively. First, a sequence of velocity measurements provides considerably more information than a single value. For example, a single velocity measurement of 60 cm/s might occur at the peak of a small wave, or during either the accelerating or decelerating phase of a large wave. By using a sequence of three velocity measurements (spread over approximately 1/4 wave period), it is possible to distinguish all of these situations. Forecasting was also able to disprove the hypothesis (for this set of data) that the concentration during one wave was controlled by previous waves, because inclusion of such a long velocity history led to poorer forecasts.

Second, a nonlinear model is able to use a different relation between flow and concentration for each of the three flow situations. Nonlinear models allow different relations to be used for accelerating and decelerating phases of a wave, as well as for waves with different peak velocities.

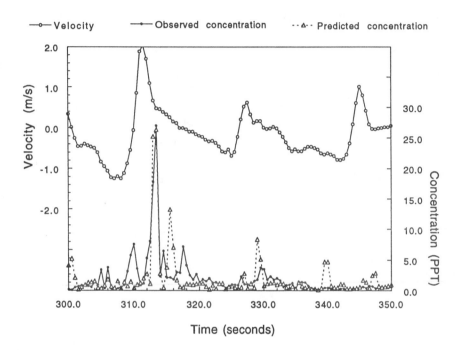

Figure 5.19: Results of input-output modeling of suspended sediment concentration in the surf zone (from Jaffe and Rubin, in press). Concentration does not mimic the flow, but nevertheless can be predicted from the flow. The best models are nonlinear (small number of nearest neighbors k), 1.5 seconds between measurements, an embedding dimension m of 3 (during a time interval of 3.0 seconds), and a lag n of 1.5 seconds. The resulting correlation coefficient between predicted and observed concentration is 0.6. For comparison, the correlation between simultaneous values of flow velocity and concentration is zero.

The third modeling requirement is a lag time between the end of each observed flow sequence and the observed concentration. Intuition might suggest that lag time should vary with height above the bed, because more time is required to mix sediment upward greater distances. Forecasting can be used to document and quantify this effect. The lag times that lead to the most accurate forecasts range from 1.5 seconds (to mix sediment to an elevation of 13 cm) to 8.5 seconds (to mix sediment to an elevation of 60 cm).

This example illustrates several uses of forecasting:

1. demonstrating that a time series has a deterministic structure (not merely due to stochastic or unpredictable processes),

2. developing a model that can make accurate forecasts, and

3. identifying the important processes that are at work, so that they can be studied, quantified, and incorporated in other models.

5.4.6   Global climate

As discussed above, most geological systems are subject to unsteady forcing, which makes it difficult to separate the behavior of the system from the behavior of the forcing in an observed time series. Hunter and Theiler (1992) developed a technique for forecasting such input-output systems and used global climate to illustrate this technique; their replicated results are discussed below.

In this example, the history of global climate (output) is represented by Imbrie et al.'s (1984) time series of $\delta^{18}O$ measured in fossil foraminifera in marine cores, and solar forcing (input) is given by the astronomically calculated insolation at 65° N. latitude in July (Berger and Loutre, 1991). The approach is to formulate a variety of models employing different values of embedding dimension m and different number of nearest neighbors k. The results indicate that climate can be predicted from the insolation (Figure 5.20), with a correlation coefficient of 0.64. To achieve this level of performance requires using a nonlinear model with an embedding dimension of 4 (during a time interval of 6000 years). The best linear model does not perform as well (correlation coefficient of 0.52). Because models based on other spectral analysis techniques are linear, they are not as accurate as the nonlinear models developed using Equation 5.7.

One interesting aspect of this input-output modeling is that it allows climate to be predicted for long times in the future, with little loss in expected accuracy. Although the relation between insolation and climate is nonlinear, insolation, which is controlled by periodic (linear) astronomical processes, can be predicted accurately for long times into the future (Berger and Loutre, 1991). Nonlinear modeling using Equation 5.7 can then be used to predict the climate from the predicted insolation.

5.4.7   Summary of uses and limitations of forecasting

The discussion and examples in this chapter illustrate a variety of applications of forecasting. The two main uses are to predict the future and to characterize systems. Making predictions (such as global climate or sediment transport) has obvious practical value; the value of system characterization is less obvious but includes:

1. Forecasting can be used to demonstrate that the irregular, nonperiodic, behavior of a system is deterministic, rather than stochastic (demonstrating that complicated behavior is caused by a simple, non-random, process).

2. Forecasting can be used to quantify a link between forcing (such as wave-generated

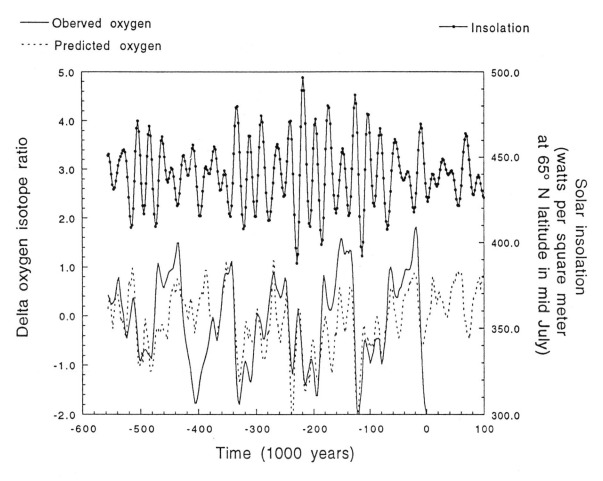

Figure 5.20: Predicted and observed climate (δ^{18}O) forecast from insolation. The best models are nonlinear and employ and embedding dimension of 4 (during a time interval of 6000 years). The correlation coefficient between predicted and observed δ^{18}O is 0.64. The correlation coefficient for the best linear model is 0.52. Insolation calculations are from (Berger and Loutre, 1991), and time is plotted relative to 1950 (negative in the past and positive in the future).

flow velocity) and system response (such as sediment transport), including the magnitude and time lag of the response to the forcing.

3. Forecasting can be used to quantify the temporal or spatial scales over which a particular process operates (for example, quantifying the temporal scale important to resuspension of sediment in the surf zone or quantifying the spatial scale over which one bedform influences another).

4. By quantifying the relations listed above, forecasting can provide guidance for more traditional modeling efforts.

Forecasting has a number of important requirements and limitations. Many of these requirements are difficult or impossible to meet in the earth sciences:

1. The time series must be stationary (similar mean and standard deviation in the testing and learning sets) and recurrent (similar events must recur).

2. Forecasting requires lengthy time series (not merely requiring many data points, but requiring data points that sample many events).

3. The structure of the time series must arise from internal behavior (or self-organization) of the system, rather than from external forcing. Alternatively, external forcing must be measured and addressed using input-output modelling.

4. Because small amounts of noise can mask nonlinearity, the time series should be relatively free of noise.

5. Because natural time series contain noise, the performance of nonlinear models is often only slightly better than linear models. Although the better performance of nonlinear models may be important for demonstrating that a system is nonlinear, the actual predictions of the nonlinear model may not be that much more accurate than linear predictions.

6. To learn from forecasting requires the formulation and testing of null hypotheses. Unless the proper null hypotheses are tested, the results of forecasting may be unimportant or misleading.

Despite the substantial limitations listed above, the concepts developed for the study of dynamical systems can help us understand geological systems. Even if an earth scientist never applies the techniques in this publication, understanding the multiple causes of complicated nonperiodic time series is essential to interpreting earth history.

References

AASUM, Y., KELKAR, M., and GUPTA, S., 1991, An application of geostatistics and fractal geometry for reservoir characterization: SPE Formation Evaluation, v. 6, p. 11–19.

ABARBANEL, H.D.I., 1992, Local and global Lyapunov exponents on a strange attractor: in M. Casdagli and S. Eubank, eds., Nonlinear Modeling and Forecasting, Reading MA, Addison-Wesley, p. 229–247.

ABARBANEL, H.D.I., in press, Analysis of Observed Chaotic Data: New York, Springer-Verlag.

ABARBANEL, H.D.I., et al., 1993, The analysis of observed chaotic data in physical systems: Reviews of Modern Physics, v. 65, no. 4, p. 1331–1392.

ABARBANEL, H.D.I., and KENNEL, M.B., 1993, Local false nearest neighbors and dynamical dimensions from observed chaotic data: Physical Reviews E, v. 47, p. 3047–3068.

ABARBANEL, H.D.I., and U. LALL, in press, Nonlinear dynamics of the Great Salt Lake: System identification and prediction: Water Resources Research.

ALLAIN, C., and CLOITRE, M., 1991, Characterizing the lacunarity of random and deterministic fractal sets: Physical Review A, v. 44, p. 3552–3558.

ALLEN, J.R.L., 1968, Current Ripples; Their Relation to Patterns of Water and Sediment Motion: Amsterdam, North-Holland Publishing Company, 433 p.

ALLEN, J.R.L., 1978, Polymodal dune assemblages: an interpretation in terms of dune creation-destruction in periodic flows: Sedimentary Geology, v. 20, p. 17–28.

ANDERSON, R.S., 1990, Eolian ripples as examples of self-organization in geomorphological systems: Earth-Science Reviews, v. 29, p. 77–96.

ARNEODO, A., GRASSEAU, G., and KOSTELICH, E.J., 1987, Fractal dimensions and $f(\alpha)$ spectrum of the Hénon attractor: Physics Letters A, v. 124, no. 8, p. 426–432.

BAK, Per, and PACZUSKI, M., 1993, Why nature is complex: Physics World, Dec. 1993, p. 39–43.

BAK, Per, and CHEN, K., 1991, Self-organized criticality: Scientific American (January 1991), p. 46–53.

BAK, Per, and CREUTZ, M., 1994, Fractals and self-organized criticality: in A. Bunde and S. Havlin, eds., Fractals in Science, New York, Springer-Verlag, p. 26–47.

BAKER, G.L., and GOLLUB, J.P., 1990, Chaotic Dynamics: An Introduction: Cambridge, Cambridge Univ. Press, 182 p.

BARTON, C., and LAPOINTE, P.R., eds., 1995a, Fractals in the Earth Sciences: New York, Plenum, 265 p.

121

122

BARTON, C., and LAPOINTE, P.R., eds., 1995b, Fractals in Petroleum Geology and Earth Processes: New York, Plenum, 317 p.

BASS, T.A., 1985, The Eudaemonic Pie: Or Why Would Anyone Play Roulette without a Computer in His Shoe? New York, Houghton-Mifflin, 324 p.

BASSINGTHWAIGHTE, J., and RAYMOND, G., 1994, Evaluating rescaled range analysis for time series: Annals of Biomedical Engineering, v. 22, p. 432–444.

BEAUMONT, C., FULLSACK, P., and HAMILTON, J., 1992, Erosional control of active compressional orogens: in K.R. McClay, ed., Thrust Tectonics, London, Chapman and Hall, p. 1–18.

BELTRAMI, H., and MARESCHAL, J.C., 1994, Strange seismic attractors: PAGEOPH, v. 141, p. 71–81.

BERAN, J., 1992, Statistical methods for data with long-range dependence: Statistical Science, v. 7, p. 404–427.

BERGÉ, P., POMEAU, Y., and VIDAL, C., 1986, Order within Chaos: Towards a Deterministic Approach to Turbulence: New York, John Wiley and Sons, translated from the 1984 French edition by L. Tucherman, 329 p.

BERGER, A., and LOUTRE, M.F., 1991, Insolation values for the climate of the last 10 million years: Quaternary Sci. Review, v. 10, p. 297–317.

BOYAJIAN, G., and LUTZ, T., 1992, Evolution of biological complexity and its relation to taxonomic longevity in the Ammonoidea: Geology, v. 20, p. 983–986.

BRIGGS, K., 1990, An improved method for estimating Lyapunov exponents of chaotic time series: Physics Letters A, v. 151, p. 27–32.

BROWN, R., BRYANT, P., and ABARBANEL, H.D.I., 1991, Computing the Lyapunov spectrum of a dynamical system from an observed time series: Physical Review A, v. 43, p. 2787–2806.

BUNDE, A., and HAVLIN, S., eds., 1991, Fractals and Disordered Systems: Berlin, Springer-Verlag, 350 p.

CAHALAN, R.F., LEIDECKER, H., and CAHALAN, G.D., 1990, Chaotic rhythms of a dripping faucet: Computers in Physics, July/August 1990, p. 368–383.

CARR, J.R., 1995, Numerical Analysis for the Geological Sciences: Englewood Cliffs, N.J., Prentice-Hall, 592 p.

CARR, J.R. and BENZER, W.B., 1991, On the practice of estimating fractal dimension: Mathematical Geology, v. 23, p. 947–958

CASDAGLI, M., 1992, Chaos and deterministic versus stochastic nonlinear modeling: Jour. Roy. Statistical Soc. B, v. 54, p. 303–328.

CASDAGLI, M., and A.S. WEIGEND, 1994, Exploring the continuum between deterministic and stochastic modeling: in Weigend, A.S., and Gershenfeld, N.A., eds., Time Series Prediction: Forecasting the Future and Understanding the Past, Reading, Massachusetts, Addison-Wesley, p. 347–366.

CHOPARD, B., HERRMANN, H., and VICSEK, T., 1991, Structure and growth of mineral dendrites: Nature, v. 353, p. 409–411.

CORTINI, M., and BARTON, C.C., 1993, Nonlinear forecasting analysis of inflation-deflation patterns of an active caldera

(Campi Flegrei, Italy): Geology, v. 21, p. 239–242.

CRUTCHFIELD, J.P., FARMER, J.D., PACKARD, N.H., and R.S. SHAW, 1986, Chaos: Scientific American, v. 254 (Dec. issue), p. 46–57.

CRUTCHFIELD, J.P., and KANEKO, K., 1987, Phenomenology of spatio-temporal chaos: in Bai-lin, H., ed., Directions in Chaos: Singapore, World Scientific, v. 1, p. 272–353.

CRUTCHFIELD, J.P., and KANEKO, K., 1988, Are attractors relevant to turbulence?: Physical Review Letters, v.60, p.2715–2718.

DAVIS, J.C., 1986, Statistics and Data Analysis in Geology, Second Edition: New York, Wiley, 646 p.

DEVANEY, R.L., 1989, An Introduction to Chaotic Dynamic Systems, Second edition: Redwood City, CA, Addison-Wesley, 336 p.

DEUTSCH, C., and JOURNEL, A., 1992, GSLIB Geostatistical Software Library and Users Guide: New York, Oxford University Press, 340 p.

DUBOIS, J., and CHEMINÉE, J.-L., 1993, Les cycles éruptifs du Piton de la Fournaise: analyse fractale, attracteurs, aspects déterministes: Bull. Geol. Soc. France, v. 164, no. 1, p. 3–16.

ELGAR, S., and KADTKE, J., 1993, Paleoclimate attractors: new data, further analysis: Internatl. Jour. Bifurcations and Chaos, v. 3, p. 1587–1590.

EMANUEL, A., ALAMEDA, G., BEHRENS, R., and HEWETT, T., 1989, Reservoir performance prediction methods based on fractal geostatistics: SPE Reservoir Engineering, August, 1989, p. 311–318.

FARMER, J.D., and SIDOROWICH, J.J., 1988, Exploiting chaos to predict the future and reduce noise, in Lee, Y.C., ed., Evolution, Learning, and Cognition: Singapore, World Scientific.

FEDER, J., 1988, Fractals: New York, Plenum, 283 p.

FIELD, R.L., and GYÖRGI, L., eds., 1993, Chaos in Chemistry and Biochemistry: Singapore, World Scientific, 289 p.

FORREST, S.B., and HAFF, P.K., 1992, Mechanics of wind ripple stratigraphy: Science, v. 255, p. 1240–1243.

FREDKIN, D.R., and RICE, J.A., 1995, Method of false nearest neighbors: a cautionary note: Physical Review E, v. 51, p 2950–2954.

FLUEGEMANN, R.H., JR., and SNOW, R.S., 1989, Fractal analysis of long-range paleoclimatic data: oxygen isotope record of Pacific core V28-239: PAGEOPH, v. 131, p. 309–313.

FORD, J., 1983, How random is a coin toss? Physics Today, April, p. 40–47.

FRASER, A.M., and SWINNEY, H.L., 1986, Independent coordinates for strange attractors from mutual information: Physical Reviews A, v. 33, p. 1134–1140.

FRITSCH, U., and ORSZAG, S.A., 1990, Turbulence: challenges for theory and experiment: Physics Today, v. 43, p. 24–32.

FRØYLAND, J., and ALFSEN, K.H., 1984, Lyapunov-exponent spectra for the Lorenz model: Physical Review A, v. 29, p. 2928–2931.

GARCIA-RUIZ, J.M., CHECA, A., and RIVAS, P. 1990, On the origin of ammonite sutures: Paleobiology, v. 16, p. 349–354.

124

GEFEN, Y., MEIR, Y., and AHARONY, A., 1983, Geometric implementation of hypercubic lattices with noninteger dimensionality by use of low lacunarity fractal lattices: Physical Review Letters, v. 50, p. 145–148.

GERSHENFELD, N.A., and WEIGEND, A.S., 1994, The future of time series: learning and understanding: in Weigend, A.S. and Gershenfeld, N.A., eds., Time Series Prediction: Forecasting the Future and Understanding the Past, Reading MA, Addison-Wesley, p. 1–70.

GODANO, C., and SALERNO, M., 1993, The chaocity degree of the Campi Flegrei seismicity, southern Italy: Geophysical Jour. International, v. 114, p. 392–398.

GOLLUB, J. P., BENSON, S. V., and STEINMAN, J., 1980, A subharmonic route to turbulent convection: in Helleman, R.H.G., ed., Nonlinear Dynamics: Annals of the New York Academy of Sciences, v. 357, p. 22–27.

GOLLUB, J. P., and SWINNEY, H.L., 1975, Onset of turbulence in a rotating fluid: Physical Review Letters, v.35, p. 927–930.

GOULD, H., and TOBOCHNIK, J., 1988, An Introduction to Computer Simulation Methods: Application to Physical Systems: Reading, MA, Addison-Wesley, 2 vols., 695 p.

GRASSBERGER, P., and PROCACCIA, I., 1983, Measing the strangeness of strange attractors: Physica D, v.9, p. 189–208.

GRASSBERGER, P., SCHREIBER, T., and SCHAFFRATH, C., 1991, Nonlinear time sequence analysis: Internatl. Jour. Bifurcations and Chaos, v. 1, p. 512–547.

GUTOWITZ, H., 1991, Cellular Automata: Theory and Experiment: Cambridge, MA, The MIT Press, 483 p.

HAKEN, H., 1985, Light, v.2, Laser Light Dynamics: New York, North-Holland, 336 p.

HARDY, H., 1992, The fractal character of photos of slabbed cores: Mathematical Geology, v. 24, p.

HASTINGS, A., et al., 1993, Chaos in ecology: Is Mother Nature a strange attractor? Annual Review Ecology Systematics, v.24, p. 1–33.

HAYES, B., 1984, The cellular automaton offers a model of the world and a world unto itself: Sci. American, v.250, March, p. 12–21.

HÉNON, M., and C. HEILES, 1964, The applicability of the third integral of motion: some numerical experiments: Astronomical Jour., v.69, p. 73–79.

HEWETT, T.A., 1986, Fractal distribution of reservoir heterogeneity and their influence on fluid transport: Society of Petroleum Engineering Paper SPE 15386.

HOLDEN, A.V., ed., 1986, Chaos: Princeton Univ. Press, 324 p.

HSUI, A., RUST, K., and KLEIN, G., 1993, A fractal analysis of Quaternary, Cenozoic-Mesozoic, and Late Pennsylvanian sea-level changes: Journal of Geophysical Research, v. 98, p. 21,963–21,967.

HUNTER, N., and THEILER, J., 1992, Nonlinear signal processing: the time series analysis of driven nonlinear systems: Los Alamos report LA-UR-92-1268, 67 p.

HURST, J.E., 1951, Long-term storage capacity of reservoirs: Transactions American Society of Civil Engineers, v. 116, p. 770–808.

ITO, K., 1980, Chaos in the Rikitake two-disk dynamo system: Earth Planetary Sci. Letters, v. 51, p. 451–456.

IMBRIE, J., HAYS, J.D., MARTINSON, D.G., MCINTYRE, A., MIX, A.C., MORLEY, J.J., PISIAS, N.G., PRELL, W.L., and SHACKLETON, N.J., 1984, The orbital theory of Pleistocene climate: support from a revised chronology of the marine $\delta^{18}O$ record: in Berger, A.L., et al., eds., Milankovitch and Climate: D. Reidel Publishing Company, p. 269–305.

JACKSON, E.A., 1989, 1990, Perspectives on Nonlinear Dyanmics: Cambridge Univ. Press, v.1, 495 p., v.2, 632 p.

JACKSON, R.G., 1976, Sedimentological and fluid dynamic implications of the turbulent bursting phenomenon in geophysical flows. Journal of Fluid Mechanics, v. 77, p. 531–560.

JAFFE, B.E., and RUBIN, D.M., in press, Using nonlinear forecasting to learn the magnitude and phasing of time-varying sediment suspension in the surf zone: Journal of Geophysical Research (Oceans).

KAANDORP, J.A., 1994, Fractal Modelling: Growth and Form in Biology: Berlin, Springer-Verlag, 208 p.

KAPLAN, D.T., and GLASS, L., 1993, Coarse-grained embeddings of time series: random walks, Gaussian random processes, and deterministic chaos: Physica D, v.64, p. 431–454.

KAUFFMAN, S.A., 1993, The Origins of Order: Self-Organization and Selection in Evolution: Oxford University Press, 709 p.

KELLOGG, L.H., 1992, Mixing in the mantle: Annual Review Earth Planetary Sci., v.20, p.365–388.

KELLOGG, L.H., and D.L. TURCOTTE, 1990, Mixing and the distribution of heterogeities in a chaotically convecting mantle: Jour. Geophysical Research, v. 95, no. B1, p. 421–432.

KENNEL, M., BROWN, R., and ABARBANEL, H., 1992, Determining embedding dimension for phase-space reconstruction using a geometric construction: Physical Reviews A, v.45, p.3403–3411.

KLEMES, V., 1974, The Hurst phenomenon: a puzzle?: Water Resources Research, v. 10, p. 675–688.

KLINKENBERG, B., and GOODCHILD, M., 1992, The fractal properties of topography: a comparison of methods: Earth Surface Processes and Landforms, v. 17, p. 217–234.

KOLÁŘ, M., and G. GUMBS, 1992, Theory for the experimental observation of chaos in a rotating waterwheel: Physical Review A, v.45, p.626–637.

KORVIN, G., 1992, Fractal Models in the Earth Sciences: Amsterdam, Elsevier, 396 p.

KROHN, C.E., 1988a, Sandstone fractal and Euclidean pore volume distributions: Journal of Geophysical Research, v. 93, p.3286–3296.

KROHN, C.E., 1988b, Fractal measurements of sandstones, shales, and carbonates. Journal of Geophysical Research, v. 93, p. 3297–3305.

LANDAU, L.D., 1944, Turbulence: Dokl. Acad. Nauk. SSSR, v.44, no. 8, p. 339–342.

LAPOINTE, P. and BARTON, C., 1995, Creating reservoir simulations with fractal characteristics, in Barton, C. and Lapointe, P.R., eds., Fractals in Petroleum Geology and Earth Processes: New York, Plenum, p. 263–278.

LAVALLÉE, D., LOVEJOY, S., SCHERTZER, D., and LADOY,

P., 1993, Nonlinear variability of landscape topography: multifractal analysis and simulation, in Lam, N. and DeCola, L. , eds., Fractals in Geography: Englewood Cliffs, N.J., PTR Prentice Hall, p. 158–192.

LENDARIS, G.G., and FRASER, A.M., 1994, Visual fitting and extrapolation: in Weigend, A.S., and Gershenfeld, N.A., eds., Time Series Prediction: Forecasting the Future and Understanding the Past, Reading, Massachusetts, Addison-Wesley, 319–322.

LI, T.Y., and YORKE, J.A., 1975, Period three implies chaos: Amer. Math. Monthly, v. 82, p. 983–992.

LORENZ, E.N., 1963, Deterministic nonperiodic flow: Jour. Atmospheric Sci., v. 20, p. 130–141.

LORENZ, E.N., 1991, Dimension of weather and climate attractors: Nature, v. 353, p. 241–244.

LORENZ, E.N., 1993, The Essence of Chaos: Seattle, Univ. Washington Press, 227 p.

LUDWIG, J., and REYNOLDS, J., 1988, Statistical Ecology: New York, Wiley, 337 p.

MALINVERNO, A., 1989, Testing linear models of sea-floor topography: PAGEOPH, v. 131, p. 139–155.

MALINVERNO, A., 1995, Fractals and ocean floor topography: a review and a model, in Barton, C. and Lapointe, P.R., eds., Fractals in the Earth Sciences: New York, Plenum, p. 107–130.

MANDELBROT, B., 1965, Self-similar error clusters in communications systems and the concept of conditional stationarity: IEEE Transactions on Communications Technology, v. 13, p. 71–90.

MANDELBROT, B., 1971, A fast fractional Gaussian noise generator: Water Resources Research, v. 7, p. 543–553.

MANDELBROT, B., 1974, Intermittent turbulence in self-similar cascades: divergence of high moments and dimension of the carrier: Journal of Fluid Mechanics, v. 62, p. 331–358.

MANDELBROT, B., 1982, The Fractal Geometry of Nature: New York, W.H. Freeman and Co, 468 p.

MANDELBROT, B., 1989, Multifractal measures, especially for the geophysicist: PAGEOPH, v. 131, p. 5–42.

MANDELBROT, B. 1995. Presentation of Mandelbrot and Wallis 1969. in Barton, C. and Lapointe, P.R., eds., 1995a. Fractals in the Earth Sciences: New York, Plenum, p. 41.

MANDELBROT, B., and VAN NESS, J., 1968, Fractional Brownian motions, fractional noises, and applications. SIAM Review, v. 10, p. 422–437.

MANDELBROT, B., and WALLIS, J.R., 1968, Noah, Joseph, and operational hydrology: Water Resources Research, v. 4, p. 909–918.

MANDELBROT, B., and WALLIS, J.R., 1969a, Some long-run properties of geophysical records: Water Resources Research, v. 5, p. 321–340.

MANDELBROT, B., and WALLIS, J.R., 1969b, Robustness of the rescaled range R/S in the measurement of long-range statistical dependency: Water Resources Research, v. 5, p. 967-988.

MAÑÉ, R., 1981, On the dimension of the compact invariant sets of certain nonlinear maps. Lecture Notes in Math 898, New York, Springer-Verlag, p. 230–242.

MARESCHAL, J.-C., 1989, Fractal reconstruction of sea floor topography: PAGEOPH, v. 131, p. 197–210.

MARSHALL, C., 1991, Estimation of taxonomic ranges from the fossil record: Short Courses in Paleontology, number 4, p. 19–38.

MAY, R.M., 1976, Simple mathematical models with very complicated dynamics: Nature, v. 261, p. 459–467.

MENEVEAU, C., and SREENIVASAN K.R., 1989, Measurement of $f(\alpha)$ from scaling of histograms and applications to dynamic systems and fully developed turbulence: Physics Letters A,, v. 137, p.103–112.

MERINO, E., 1992, Self-organization in stylolites: American Scientist, v. 80, p. 466–473.

MESA, O., and POVEDA, G., 1993, The Hurst effect: the scale of fluctuation approach: Water Resources Research, v. 29, p. 3995–4002.

MIDDLETON, G.V., and WILCOCK, P.R., 1994, Mechanics in the Earth and Environmental Sciences: Cambridge, Cambridge Univ. Press, 459 p.

MIDDLETON, G.V., and SOUTHARD, J.B., 1984, Mechanics of Sediment Movement: Tulsa, Oklahoma, Society of Economic Paleontologists and Mineralogists Short Course no. 3, 401 p.

MOLLO-CHRISTENSEN, E., 1973, Intermittency in large scale turbulent flows: Annual Review of Fluid Mechanics, v. 5, p. 101–118.

MOON, F.C., 1992, Chaotic and Fractal Dynamics: An Introduction for Applied Scientists and Engineers: New York, John Wiley and Sons, Inc., 508 p.

MUDELSEE, M. and STATTEGGER, K., 1994, Application of the Grassberger-Procaccia algorithm to the $\delta^{18}O$ record from ODP site 659: selected methodical aspects: in Kruhl, J.H., ed., Fractals and Dynamic Systems in Geoscience. New York, Springer-Verlag, p. 399–413.

MUZY, J.F., BACRY, E., and ARNEODO, A., 1994, The multifractal formalism revisited with wavelets: Internatl. Jour. Bifurcations and Chaos, v. 4, p. 245–302.

NEWLAND, D.E., 1993, Random Vibrations, Spectral and Wavelet Analysis: 3rd edition, Harlow, U.K., Longman Scientific and Technical, 477 p.

NEWMAN, W.I., GABRIELOV, A., and TURCOTTE, D.L., eds., 1994, Nonlinear Dynamics and Predictability of Geophysical Phenomena: Amer. Geophysical Union, Geoph. Mon. 83, 107 p.

NICOLIS, G., and PRIGOGINE, I., 1977, Self-Organization in Nonequilibrium Systems. From Dissipative Systems to Order through Fluctuations: New York, John Wiley and Sons, 491 p.

NORDIN, C., MCQUIVEY, R., and MEJIA, J., 1972, Hurst phenomenon in turbulence: Water Resources Research, v. 8, p. 1480–1486.

NOWELL, A.R., 1978, Dissipation and fine-scale structure in turbulent open channel flow: Water Resources Research, v. 14, p. 517–526.

OLSEN, L.F., and DEGN, H., 1985, Chaos in biological systems: Quarterly Review of Biophysics, v. 18, p. 165–225.

ORTOLEVA, P.J., ed., 1990, Self-organization in geological systems: Earth-Science Reviews, v. 29, nos. 1–4.

128

ORTOLEVA, P., MERINO, E., MOORE, C., and CHADAM, J., 1987, Geochemical self-organization I: reaction-transport feedbacks and modeling approach: American Jour. Sci., v. 287, p. 979–1007.

OTT, E., 1993, Chaos in Dynamical Systems: Cambridge Univ. Press, 385 p.

OTT, E., SAUER, T., and YORKE, J.A., eds., 1994, Coping with Chaos: Analysis of Chaotic Data and the Exploitation of Chaotic Systems: New York, John Wiley and Sons, 418 p.

PACKARD, N.H., CRUTCHFIELD, J.P., FARMER, J.D., and SHAW, R.S, 1980, Geometry from a time series: Physics Review Letters, v. 45, p. 712–716.

PALMER, T.N., 1993, Extended-range atmospheric prediction and the Lorenz model: Bull. Amer. Meteorological Soc., v. 74, p. 49–65.

PAOLA, C., and BORGMAN, L., 1991, Reconstructing random topography from preserved stratification: Sedimentology, v. 38, p. 553–565.

PARKER, T.S. and L.O. CHUA, 1990, Practical Numerical Algorithms for Chaotic Systems: New York, Springer-Verlag, 348 p.

PEITGEN, H-O., H. JÜRGENS, and D. SAUPE, 1992, Chaos and Fractals: New Frontiers of Science: New York, Springer-Verlag, 984 p.

PEITGEN, H.-O., and RICHTER, P.H., 1986, The Beauty of Fractals: Berlin, Springer-Verlag, 199 p.

PLOTNICK, R., 1986, A fractal model for the distribution of stratigraphic hiatuses: Journal of Geology, v. 94, p.885–890.

PLOTNICK, R., GARDNER, R., and O'NEILL, R., 1993, Lacunarity indices as measures of landscape texture. Landscape Ecology, v. 8, p. 201–211.

PLOTNICK., R., and PRESTEGAARD, K., 1993, Fractal analysis of geologic time series: in Lam, N. and DeCola, L. , eds., Fractals in Geography, Englewood Cliffs, N.J., PTR Prentice Hall, p. 193–210.

PLOTNICK., R., and PRESTEGAARD, K., 1995, Fractal and multifractal models and methods in stratigraphy: in Barton, C. and Lapointe, P.R., eds., Fractals in Petroleum Geology and Earth Processes, New York, Plenum, p. 73–96.

PRESS, W.H., FLANNERY, B.P., TEUKOLSKY, S.A., and VETTERING, W.T., 1989, Numerical Recipes in Pascal: The Art of Scientific Computing: Cambridge Univ. Press, 759 p.

PRESTEGAARD, K., and PLOTNICK, R. 1995, New models require new data: fractal and multifractal measures of bedload transport: in Barton, C. and Lapointe, P.R., eds., Fractals in Petroleum Geology and Earth Processes, New York, Plenum, p. 113–126.

PRUSINKIEWICZ, P., and LINDENMEYER, A., 1990, The Algorithmic Beauty of Plants: Berlin, Springer-Verlag, 228 p.

RAUDKIVI, A.J., 1963, A study of sediment ripple formation: Amer. Soc. Civil Eengineers, Jour. Hydraulics Division, v.89, no. 6, p. 15–33.

RIGON, R., RINALDO, A., and RODRIGUEZ-ITURBE, I., 1994, On landscape self-organization: Jour. Geophysical Research, v. 99, no. B6, p. 11,971–11,993.

RINALDO, A., et al., 1993, Self-organized fractal river networks: Phys. Rev. Lett., v. 70, no. 6, p. 822–825.

ROBERT, A., 1991, Fractal properties of simulated bed profiles in coarse-grained channels: Mathematical Geology, v. 23, p. 367–382.

ROSENSTEIN, M.T., COLLINS, J.J. and DE-LUCA, C.J., 1993, A practical method for calculating the largest Lyapunov exponents from small data sets: Physica D, v. 65, p. 117–134.

RUBIN, D.M., 1987, Cross-Bedding, Bedforms, and Paleocurrents: Tulsa, Oklahoma, Society of Economic Paleontologists and Mineralogists, 187 p.

RUBIN, D.M., 1992, Use of forecasting signatures to help distinguish periodicity, randomness, and chaos in ripples and other spatial patterns: Chaos, v. 2, p. 525–535.

RUBIN, D.M., and MCDONALD, R.R., 1995, Nonperiodic eddy pulsations: Water Resources Research, v. 31, p. 1595–1605.

RUELLE, D., 1979, Sensitive dependence on initial conditions and turbulent behavior of dynamical systems: Annals New York Acad. Sci., v. 316, p. 408–416.

RUELLE, D., 1994, Where can one hope to profitably apply the ideas of chaos? Physics Today, v. 47, no. 7 (July), p. 24–30.

RUELLE, D., and TAKENS, F., 1971, On the nature of turbulence: Communications Math. Phys., v. 20, p. 167–192.

RUSS, J., 1994. Fractal Surfaces: New York, Plenum, 309 p.

SADLER, P. M., 1981, Sediment accumulation rates and the completeness of stratigraphic sections: Journal of Geology, v. 89, p. 569–584.

SADLER, P.M., and STRAUSS., D.J., 1990, Estimation of completeness of stratigraphical sections using empirical data and theoretical models: Journal of the Geological Society of London, v. 147, p. 471–485.

SAUER, T., YORKE, J.A., and CASDAGLI, M., 1991, Embedology: Jour. Statistical Physics, v. 65, p. 579–616.

SAUPE, D., 1988, Algorithms for random fractals: in Peitgen, H. and Saupe, D., eds.. The Science of Fractal Images, New York, Springer-Verlag, p. 71–136.

SCHINDEL, D.E., 1980, Microstratigraphic sampling and the limits of paleontological resolution: Paleobiology, v. 6, p. 408–426.

SCHROEDER, M., 1991, Fractals, Chaos, Power Laws: New York, W.H. Freeman, 429 p.

SCOTT, G.K., 1994, Chemical Chaos: Oxford Univ. Press, 480 p.

SMITH, R., 1991, The application of cellular automata to the erosion of landforms: Earth Surface Processes and Landforms, v. 16, p. 273–281.

SNOW, S., 1989, Fractal sinuosity of stream channels: PAGEOPH, v. 131, p. 99–109.

SOUTHARD, J.B., and DINGLER, J.R., 1971, Flume study of ripple propagation behind mounds on flat sand beds: Sedimentology, v. 16, p. 251–263.

SPROTT, J.C., 1993, Strange Attractors: Creating Patterns in Chaos: New York, M & T Books, 426 p.

SPROTT, J.C., 1994, Some simple chaotic flows: Physical Reviews E, v. 50, p. R647–R650.

STANLEY, H.E. 1991. Fractals and multifractals: the interplay of physics and geometry, in Bunde, A. and Havlin, S., eds., 1991, Fractals and Disordered Systems: Berlin, Springer-Verlag, p. 1–49.

STOKER, J.J., 1950, Nonlinear Vibrations in Mechanical and Electrical Systems: New York, Interscience, 273 p.

STRAUSS, D., and SADLER, P.M., 1989, Classical confidence intervals and Bayesian probability estimates for ends of local taxon ranges: Mathematical Geology, v. 21, p. 411–427.

STROGATZ, S.H., 1994, Nonlinear Dynamics and Chaos, with Applications to Physics, Biology, Chemistry and Engineering: Reading, MA, Addison-Wesley Publ. Co., 498 p.

SUGIHARA, G., 1994, Nonlinear forecasting for the classification of natural time series: Phil. Trans. Royal Society of London, v. A348 (1688), p. 477–495.

SUGIHARA, G., and MAY, R.M., 1990, Nonlinear forecasting as a way of distinguishing chaos from measurement error in time series: Nature, v. 344, p. 734–741.

TAKENS, F., 1981, Detecting strange attractors in turbulence: in D.A. Rand and L.-S. Young, eds., Dynamical Systems and Turbulence, Lecture Notes in Math. 898, New York, Springer-Verlag, p. 366–381.

THEILER, J., 1986, Spurious dimension from correlation algorithm applied to limited time series data: Physical Review A, v. 34, p. 2427–2432.

THEILER, J., EUBANK, S., LONGTIN, A., GALDRIKIAN, B., and FARMER, J.D., 1992, Testing for nonlinearity in time series: the method of surrogate data: Physica D, v. 58, p. 77–94.

THEILER, J., GALDRIKIAN, B., LONGTIN, A., EUBANK, S., and FARMER, J.D., 1992, Using surrogate data to detect nonlinearity in time series, in Casdagli, M. and Eubank, S., eds., Nonlinear Modeling and Forecasting: Reading, Massachusetts, Addison-Wesley, p. 163–188.

THEILER, J., LINSAY, P.S., and RUBIN, D.M., 1994, Detecting nonlinearity in data with long coherence times: in Weigend, A.S., and Gershenfeld, N.A., eds., Time Series Prediction: Forecasting the Future and Understanding the Past, Reading, Massachusetts, Addison-Wesley, p. 429–455.

THOMPSON, A.H., 1991. Fractals in rock physics: Annual Review of Earth and Planetary Sciences, v. 19, p. 237–262.

THOMPSON, J.M.T., and STEWART, H.B., 1986, Nonlinear Dynamics and Chaos: Geometrical Methods for Engineers and Scientists: New York, John Wiley and Sons, 376 p.

THORNE, J. 1995, On the scale independent shape of prograding stratigraphic units: applications to sequence stratigraphy: in Barton, C. and Lapointe, P.R., eds., Fractals in Petroleum Geology and Earth Processes, New York, Plenum, p. 97–112.

TOFFOLI, T., and MARGOLIS, N., 1987, Cellular Automata Machines: A New Environment for Modeling: Cambridge, MA, The MIT Press, 259 p.

TRITTON, D.J., 1988, Physical Fluid Dynamics, Second Edition: Oxford, Clarendon Press, 519 p.

TSONIS, A.A., 1992, Chaos, from Theory to Applications: New York, Plenum Press, 274 p.

TUBMAN, K.M., and CRANE, S.D., 1995. Vertical vs. horizontal well log variability and application to fractal reservoir modeling: in Barton, C. and Lapointe, P.R., eds.,

Fractals in Petroleum Geology and Earth Processes, New York, Plenum, p. 279–294.

TURCOTTE, D.L., 1992, Fractals and Chaos in Geology and Geophysics: Cambridge, Cambridge University Press, 221 p.

TURCOTTE, D.L., 1994a, Fractal aspects of geomorphic and stratigraphic processes: GSA Today, v. 4, p. 201, 211–213.

TURCOTTE, D.L., 1994b, Fractal theory and the estimation of extreme floods: Journal of Research of the National Institute of Standards and Technology, v. 99, p. 377–389.

UEDA, Y., 1980, Steady motions exhibited by Duffing's equation: A picture book of regular and chaotic motions: in P.J. Holmes, ed., New Approaches to Nonlinear Problems in Dynamics, Philadelphia, SIAM, p. 311–322.

VOSS, R., 1988. Fractals in nature: from characterization to simulation: in Peitgen, H. and Saupe, D., eds.. The Science of Fractal Images, New York, Springer-Verlag, p. 22–70.

WALLIS, J., and MATALAS, N., 1969, Small sample properties of H and K-estimators of Hurst coefficient H: Water Resources Research, v. 5, p. 1583–1594.

WEIGEND, A.S., and GERSHENFELD, N.A., eds., 1994, Time Series Prediction: Forecasting the Future and Understanding the Past: Reading MA, Addison-Wesley, 643 p.

WERNER, B.T., and FINK, T.M., 1993, Beach cusps as self-organized patterns: Science, v. 260, p. 968–971.

WHITNEY, H., 1936, Differentiable manifolds: Annals of Mathematics, v. 37, p. 645–680.

WOLF, A., SWIFT, J.B., SWINNEY, H.L., and VASTANO, J.A., 1985, Determining Lyapunov exponents from a time series: Physica D, v. 16, p. 285–317.

WONG, P., 1988, The statistical physics of sedimentary rock: Physics Today, v. 41, p. 24–32.

YUEN, D.A., ed., 1992, Chaotic Processes in the Geological Sciences: New York, Springer-Verlag, 317 p.

ZENG, X., PIELKE, R.A., and EYKHOLT, R., 1993, Chaos theory and its application to the atmosphere: Bull. Amer. Meteorological Soc., v. 74, p. 631–644.

Appendix I: Chaossary
A Short Glossary of Chaos

Gerard V. Middleton
Department of Geology
McMaster University
Hamilton ON L8S 4M1, Canada

This is an expanded and revised version of the glossary that appeared in Middleton (1991). Phrases in bold-face have separate entries in the glossary. Other useful glossaries are found in Jackson (1989), Moon (1992) and Thompson and Stewart (1993).

Aggregation

When two particles collide and stick together (without completely merging) they are said to aggregate. A particular form of aggregation that has been much studied is **diffusion-limited aggregation** (**DLA**: first named by Witten and Sander in 1981, see Sander, 1987). In DLA a seed particle grows by aggregation of other particles approaching it by a random-walk diffusive process. The aggregates that result have **fractal** properties and resemble certain natural dendrites: their study has also thrown light on the formation of skeletal crystals (such as snow flakes). Similar structures are formed by quite different physical processes, such as viscous fingering in porous media, and chemical dissolution.

Arnol'd Tongues

see **circle map**.

Attractor

The term is used quite generally to describe the state towards which dissipative dynamic systems (and their mathematical analogues) tend to converge. A pendulum with friction, for example, tends to converge on the position of rest, with the bob hanging vertically below the support. A pendulum with friction which is periodically given an impulse may tend to converge on a particular cycle of motion, called a **limit cycle** (a simple equation to describe such a pendulum or oscillator is called the **van de Pol equation**, and has application in electronics). These types of attractors have been known for many years, but a new type of attractor, now called a **strange attractor**, was described for the first time by a meteorologist, Lorenz, in 1963. Lorenz, however, did not use the term itself. It is not clear who did originate the term: several authors attribute it to Smale, 1967, but though he recognized many of the properties of strange attractors, and certainly refered to "attractors" in his 1967 paper, the term "strange attractors" does not seem to appear in the paper. Ruelle, 1980, writes that he asked Takens if he had originated it, in Ruelle and Takens (1971), and he replied "Did you ever ask God if he created this damned universe...I don't remember anything..."

Attractors are generally portrayed either in **phase** (or **state**) **space** or in planes cut through phase space. These planes are called **Poincaré sections**.

Strange Attractors are characteristic of certain dissipative dynamic systems of more that two dimensions, and describe a particular kind of chaotic behaviour. Their most characteristic feature is that, although the attractor is confined within a certain region of phase space, trajectories within the attractor show "sensitivity to initial conditions". This means that trajectories that begin from positions very close together diverge very rapidly (in fact, exponentially) so that their positions in phase space are soon very far apart. A measure of the rate of divergence is provided by the **Lyapunov exponent**. Divergence takes place even though both trajectories are constrained to lie within, or very close to, the attractor. Divergence of trajectories may be thought of as a stretching process, because a group of points originally confined within a small cube, for example, commonly become "stretched out" and are confined within a thin sheet-like volume down trajectory. Divergence of trajectories may be reconciled with confinement, by the process of "folding" trajectories back on themselves in phase space.

Most (but perhaps not all) strange attractors show fractal properties. Trajectories along the surface of an attractor are smooth, so a single trajectory does not show fractal properties: it is the geometry of the surface locally normal to the trajectory that is fractal. Therefore, the fractal nature of the attractor is often most clearly shown by a Poincaré section.

Not all authors agree on what makes an attractor "strange" — therefore some have suggested that attractors should be described as "chaotic" when they show exponential divergence of nearby trajectories, and "fractal" when they show fractal geometry.

Basin of Attraction

In dissipative dynamic systems, trajectories in **phase** (or **state**) **space** tend to converge on one or more attractors. In systems with more than one attractor, it is possible to map out (theoretically, or more usually by computation)

which initial conditions (regions in phase space) lead to which attractors. These regions constitute the basin of attraction of a particular attractor. The boundaries between basins have extremely complex geometry in many dynamic systems, and constitute very striking fractals. A well-known example is the boundary between trajectories that diverge to infinity and those that converge to zero in the map

$$z^2_{n+1} = z^2_n - c$$

where z and c are complex numbers. If c is a constant, and the initial value, z_0, is plotted on the screen (using different colours or shades depending on the rate of convergence or divergence) then the boundary is called a *Julia set*. If the initial point is fixed at $z_0 = 0$, and c is plotted on the screen (using different colors for different rates of divergence) then the boundary is called the *Mandelbrot set*. Though the beauty of these sets has contributed mightily to popularizing fractals, they have no scientific applications (known to us).

Bifurcation

The general phenomena being described is the transition from a single-valued function to a double-valued function. For example, the well-known **logistic map**:

$$x_{n+1} = kx_n(1 - x_n)$$

is a single valued function for values of $k < 3$, but becomes double valued in the range $3 < k < 3.5$. At higher values of k, further bifurcations are observed, and the function becomes 4-valued, 8-valued, etc. This is an example of period doubling and of one particular type of bifurcation of interest to students of chaos, called a *pitchfork* (or *period-doubling*) bifurcation. According to Arnol'd (1986, p.2) bifurcation "...is used in a broad sense for designating all sorts of qualitative reorganizations or metamorphoses of various entities resulting from a change of the parameters on which they depend."

A general classification of bifurcations is possible (e.g., Bergé et al., 1984, Appendix A; Thompson and Stewart, 1986, Chapter 7). The topic is closely related to the classification of "catastrophes" described by catastrophe theory. The following simplified discussion is taken from Glass and Mackey (1988).

The number and/or stability of steady states or cycles which a system displays may depend on the value of one or more parameters (in the example above, the parameter k). Any such change is described as a bifurcation and the critical value of the parameter(s) defines a bifurcation point. To understand the stability of a particular state (point in the system), it is useful to consider a small region around that point. If this region is small enough, it is possible to approximate the non-linear equations governing the system by a set of linear equations. The stability of sets of linear equations is well understood and may be found from the properties of the **Jacobian** of the system. For steady states in two dimensions, the possible conditions of stability may be described (after Poincaré) as one of the following: (i) focus – paths spiral in towards the steady state; (ii) node – paths move more directly in towards the steady state; (iii) saddle point – some paths move towards and some away from the steady state. These different conditions can be recognized by examining the **eigenvalues** of the Jacobian. Forms of instability, or bifurcations, can also be recognized in this way. For example, if the system is characterised by two complex eigenvalues, then the *Hopf Bifurcation* takes place as the eigenvalues cross the imaginary axis, i.e., as the real part of the complex eigenvalues changes from negative to positive values. Such bifurcations may be further classified as *supercritical*, if a single steady state is relaced by a **limit cycle** whose amplitude gradually grows larger, or *subcritical* if the single steady state is suddenly relaced by a large amplitude limit cycle.

See also **Feigenbaum constant**.

Brusselator

A term applied informally to a system of differential equations that results from a particular set of chemical reactions:

$$
\begin{aligned}
A &\Rightarrow X \\
2X + Y &\Rightarrow 3X \\
B + X &\Rightarrow Y + D \\
X &\Rightarrow E
\end{aligned}
$$

It is discussed by Prigogine (1980) and is named for the "Brussels school" of thermodynamics, which he leads. The system gives rise to a **limit cycle**.

Cantor Dust (or Set)

A fractal produced by subdividing a line into parts, deleting some of them, and repeating the process indefinitely. Usually the line is divided into three equal parts, and the central part is deleted: in this case the fractal dimension is 0.63...

Catastrophe

The term "catastrophe theory" was introduced by E.C. Zeeman, to describe a theory originally proposed by the French mathematician, Rene Thom. "Catastrophes are abrupt changes arising as a sudden response of a system to a smooth change in external conditions." (Arnol'd, 1986, p.2). An elementary discussion is given by Zeeman (1976). Catastrophe theorists recognize seven "elementary" (types of) catastrophes. The theory is disavowed by many chaos theorists (e.g., Arnol'd, Smale) because of the highly speculative, qualitative applications made by Zeeman and his school to the social sciences.

Chaos

Some authorities doubt that a satisfactory definition of chaos can be given in the present (?chaotic) state of knowledge. The word comes from the Greek meaning empty space or void, and is commonly used to imply the absence of any order or structure. Its first use in the modern sense is perhaps that by Li and Yorke (1975).

The present use of the term in mathematics and physics is closely connected with the behaviour observed in certain relatively simple, non-linear, dynamic systems (or sets of ordinary differential equations). Under some conditions, such equations not only display instability, but unstable behaviour of a particularly complex kind. Probably the single most important feature of these systems is that "precise knowledge of the past evolution of a system over an arbitrarily long time does not aid in predicting its subsequent evolution past a limited time range" (Bergé et al., 1986, p.103). Strogatz (1994) adds to this "sensitive dependence on initial conditions", the further requirements that chaotic systems must be deterministic and show aperiodic long-term behavior. Some chaotic, dissipative systems of three or more dimensions also display the phenomenon of **strange attractors**. Not only is chaos difficult to define, it is also difficult to prove mathematically that a given system is truly chaotic. For a review and the latest proof that the Lorenz system is, in fact, chaotic see Hastings (1994).

A focus of current research is how to distinguish chaos from "**noise**" generated by a **random** or **stochastic** process. In the real world, much "noise" is probably itself generated by many-dimensional dynamic processes—though there are probably also phenomena that closely approach true random processes, where there is no correlation between one event and the next (e.g., radioactive decay). See the discussion by Ford (1983).

Circle Map

A map of the form

$$\theta_{n+1} = \theta_n + \Omega - K\ g(\theta_n)$$

where θ is a circular measure of period one (so varies from 0 to 1 rather than the usual 2π radians), and Ω (omega) is the frequency in the absence of the nonlinear term $g(\theta)$. The nonlinear term, for example, might be a sine function (giving the *Sine Circle Map*). K is a coefficient which determines the strength of the nonlinear component. Thus circle maps are a whole class of nonlinear difference equations that map a point on a circle to another point on the circle. The average number of rotations per iteration is called the *winding number* : if there is no nonlinear term ($K = 0$) then the winding number is simply equal to Ω.

Circle maps can be used to represent the dynamics of a system with two frequencies, for example, a damped pendulum with a natural frequency (normalized to one) forced by a mechanism whose driving frequency $(1 + f)$ differs from the natural frequency. At small values of f, the motion is quasi-periodic. At larger values, phase locking occurs, and the motion is periodic. As the strength of the nonlinear component increases, the system may become chaotic.

The behaviour of the oscillating system can be portrayed by plotting K against Ω. For a given K, there will be some values of Ω for which there is phase locking between the two frequencies: for these values there is resonance even if the initial ratio between the two frequencies is not exactly rational. The range of Ω values for which this is possible increases as K increases. Lines can be drawn on the diagram separating these regions from regions in which phase locking does not take place: the phase-locked regions are called **Arnol'd tongues**.

See also **Duffing's equation**.

Devil's Staircase

A name given to a continuous "curve" composed of an infinite number of steps. A simple example can be constructed from a **Cantor dust** construction. Cantor dust is produced by subdividing a line, removing part, and repeating the process indefinitely. If a bar is substituted for the line, then the process can be modified, for example, by requiring that the mass of the original bar is conserved as parts are removed: this is achieved by proportionately increasing the mass of the remaining parts. Suppose the original bar extended from $x = 0$ to $x = 1$, then a graph showing the way the cumulative mass of the pieces, $M(x)$ as a function of x, would be a Devil's Staircase (Feder, 1988, p.69). The process of concentrating an originally uniform mass distribution onto many small regions with a fractal distribution has been called *curdling* by Mandelbrot.

The Devil's Staircase may be used to represent the dynamics of driven oscillators, and similar dynamic systems (Bak, 1986). In this case the "flat" surface of the "steps" represent rational ratios between the frequencies of the oscillator and its driver produced by phase locking (resonance between the natural frequency of the oscillator and the initial frequency of its driver).

Difference Equation

An equation that relates the value of a function $f(x, t_{i+1})$ to the value of the function at an ealier time t_i, where the difference between the two times is finite. Difference equations may be studied in their own right, and arise naturally in certain branches of science (for example, in studies of sucessive populations which arise from a regular birth/death cycle) but generally they result from recasting differential equations in finite difference form, so that they may be solved numerically.

Difference equations show more chaotic behaviour than their corresponding differential equation. An example of this is the **logistic equation**, which in its differential form has an exact integral (and is therefore non-chaotic) whereas in its difference form it is the classic example of a simple one-dimensional map showing chaos.

Dimension

The common concept of dimension is the *topological dimension*, that is, the dimension of the space needed to represent an object. For a point this is zero, for a line one, for a plane 2 and for a sphere or cube 3. It is always an integer. The **embedding dimension** is a special type of topological dimension.

A second concept is that of *fractal dimension*. Various definitions of this are available: see **fractal** and **Koch curve**. A common one is the *Hausdorf-Besicovitch* dimension, which can be defined as follows: suppose we have a set of points in an n-dimensional space. We try to cover the points by a set of N hyperspheres, each of linear dimension ϵ, where N is the smallest possible number of hyperspheres necessary to cover all the points. Then the H-B dimension D is defined by:

$$D = \lim_{\epsilon \to 0} \left(\frac{\ln N(\epsilon)}{\ln(1/\epsilon)} \right)$$

The Hausdorff dimension of a set is obtained from covering that set minimally with with open balls (intervals in one dimension) with radii that may be different. The *limit capacity* involves the same process except that one is restricted to balls of equal radii. In both cases the maximum radius is allowed to go to zero yielding a measure for the set by summing over the ball diameters raised to the power D. The greatest lower bound of the powers (D) causing that measure to be zero defines the Hausdorff dimension

138

or the limit capacity, as the case may be (Essex and Nerenberg, 1990).

This definition, though rigorous mathematically, is not very useful in practice. In practice, dimensions are generally calculated by box-counting, or in various indirect ways. This is one reason for the large number of different definitions.

The term dimension, originally introduced to describe a property of sets, has been extended by many authors to describe properties of **measures**. Mandelbrot does not approve of this extension, and prefers that such properties be given a different name. See **Multifractal**.

Duffing's Equation

A differential equation for a nonlinear driven oscillator, originally developed for applications in electronics. The nonlinear term is cubic (see **oscillator**). It shows a wide range of chaotic behaviour (e.g., see Holden, 1986).

Eigenvalue and Eigenvector

Eigenvalues are *characteristic values* of some mathematical expression. The term is generally applied to the characteristic roots of a square matrix. If the matrix is $\mathbf{A}$, the eigenvalues λ are the roots of the polynomial

$$|\mathbf{A} - \mathbf{I}\lambda| = 0$$

where $\mathbf{I}$ is the unit matrix, and the vertical lines indicate the determinant of the matrix between them. Eigenvalues have many applications, from factor analysis in statistics to determining the relative magnitudes of the principal stresses or strains in rock mechanics or structural geology. They are used to determine the stability of linear differential equations (see **Jacobian**), and in the classification of different types of **bifurcations**.

The eigenvectors are vectors in space associated with each eigenvalue. In the case of a stress matrix, the eigenvectors indicate the direction of the principal stresses, and the eigenvalues their relative magnitude.

Embedding Dimension

This is a term used in the analysis of **time series**. A time series generated by a nonlinear dynamical process represents a set of observations made on a single variable, say $x(t)$. At first, it would seem to be impossible to determine from such one-dimensional data, the dimensionality of the system that is giving rise to it. But this is, in fact, possible in many cases. The technique was first applied in practice (without proof) by a group of researchers at University of California at Santa Cruz (Packard et al., 1980) so it is sometimes called the "Santa Cruz Conjecture". Some proofs were supplied the following year by Takens, and it turns out that aspects of the subject had been worked on in 1936 by the mathematician H. Whitney. A complete mathematical review is given by Sauer et al. (1991).

The technique generally used is to generate from the one-dimensional time series, a new d-dimensional time series, by using time-lags Δt. This is called "embedding the series in d-dimensional space". For example, if $d = 3$, the new time series would consist of the vectors $(x_1(t), x_1(t+\Delta t), x_1(t+2\Delta t)), (x_2(t), x_2(t+\Delta t), x_2(t + 2\Delta t)), \ldots$ In this case, one can plot the trajectory of this new series in 3-dimensional space. If this is done, for example, for a series generated from the Lorenz equations, one sees an object that closely resembles the Lorenz attractor.

In turns out that if d is large enough, the embedding restores all the essential topology of the original attractor (though of course the scaling is not preserved). The criterion usually given for "large enough" is

$$d > 2d_A$$

where d_A is the dimension of the attractor. An heuristic explanation of this inequality has been given by Henry Abarbanel (personal communication): Suppose we have a space of 3 dimensions, then two surfaces (each of dimension 2) intersect in a line (of dimension 1). This may be generalized to the intersection of any two **manifolds** (lines, surfaces, hypersurfaces), of dimension d_{m1}, d_{m2} in a space of dimension d. The dimension of the intersection is given by d_i

$$d_i = d_{m1} + d_{m2} - d$$

So the intersection of a single manifold with itself may be expected to have dimension

$$d_i = 2d_m - d$$

Now it is a property of strange attractors that the trajectories do not intersect, and so the preservation of their true topology is possible only in a space which is large enough so that $d_i < 0$. This requires $d > 2d_A$. In practice, one can often achieve a satisfactory embedding with $d > d_A$. For example, the **Lorenz attractor**, with $d_A \sim 2.06$ can be reconstructed by embedding a Lorenz time series in a 3-dimensional space.

Ergodic Hypothesis

This is a phrase commonly applied to **time series**: it supposes an equality between "ensemble" averages (the average of all possible time series) and averages, over time, of a single time series. The phrase is also used in a more general sense in thermodynamics. A simple statement given by Maxwell (see Prigogine, 1980) is: "The only assumption which is necessary for a direct proof of the problem of thermodynamic equilibrium is that the system, if left to itself, will sooner or later pass through every phase which is consistent with the equation of energy." The ergodic hypothesis about random systems with many degrees of freedom may be contrasted with the usual hypothesis in classical mechanics, that simple systems whose initial conditions and equations of motion are fully

known, will exhibit only completely predictable behaviour, with trajectories confined to a small part of phase space.

It has now been shown (see **KAM theorem**) that many systems are intermediate in character: they show well-defined regions of predictable deterministic behaviour, and other regions within which there is chaotic behaviour.

Expectation

This has a technical meaning in statistics. The expected value of a function of a continuous variable, $g(x)$ is defined as

$$E(g(x)) = \int g(x)p(x)dx$$

where $p(x)$ is the probability density function of x. If $g(x) = x$ then the expectation of x, otherwise called the expected value of x, correspondance to the average value of x. If x is a discrete rather than continuous variable, then we replace the integral sign with a sum, and we can think of the probabilities as being the "weights" associated with each value of x (or with each value of the function $g(x)$).

Feigenbaum Constant

In the sequence of **bifurcations** observed in the **logistic equation** Feigenbaum noted in 1975 that bifurcations took place at values of k of 3.000, 3.449..., 3.544..., etc., a sequence of values which appear to converge on a limit 3.569946...He further noted that the value of k at the nth bifurcation is given by

$$k_n = 3.569946\ldots - 2.6327(4.669202\ldots)^{-n}$$

A similar phenomenon is noted in examples of period doubling bifurcations observed in many other many other iterative maps (see **Poincaré section**). Though the other numbers vary from map to map the value $F = 4.669202\ldots$ does not: it is therefore a *fundamental constant* similar to π, e, etc. and it is now called the Feigenbaum constant. A simplified "explanation" of this constant is given in Holden (1986, p.50-51). Feigenbaum has shown that the universal properties of bifurcations are a consequence of **renormalization group theory**.

Fixed Point

Also called an *equilibrium point*, or a *steady state*. The term *singular point* is also used by some authors (Edelstein-Keshet, 1988, p.176). A point in phase space for which

$$x = f(x),$$

in other words, the solution of the (set of) differential equation(s) at this point (value of x) is the value of x itself (where x may be a vector). The concept is easier to understand for a difference equation

$$x_{n+1} = f(x_n),$$

where it implies that $x(n + 1) = x(n)$, i.e., iteration does not change the value of x. If the iteration corresponds to taking the next intersection of a trajectory with a Poincaré section of a phase space, then the fixed point corresponds to a limit cycle in the phase space.

Fractal

This term was introduced by Mandelbrot in 1975 and has been defined (even by him) in different ways. One version given by Feder (1988, and attributed by him to Mandelbrot) is

A fractal is a shape made of parts similar to the whole in some way.

Fractals are related to chaos by the fact that many chaotic systems display iterative mappings (see **Poincaré section**) which have fractal properties. This means that small parts of the mapping, when enlarged, closely resemble the parts of the original mapping (at the larger scale) and that this process can be continued over several scales of enlargement. Many **strange attractors** have fractal dimension.

An important property characterizing fractals is their **dimension**. Various different definitions of fractal dimension have been proposed, with the most common ones being the similarity dimension, the box-counting dimension, and the Hausdorff (or Hausdorff-Besicovitch) dimension. Rigorous definitions can be found in Feder (1988) or Mandelbrot (1982). The following heuristic introduction is taken from Sorensen (1984).

A line has one dimension and a flat plane two dimensions, and so on (this is the *topological dimension*, and is also the dimension of the space necessary to display a line or a plane—the **embedding dimension**). An irregularly shaped line may be considered to have a dimension somewhere between 1 and 2: the dimension increases as the line becomes more irregular, and only reaches 2 when the line is so irregular that it fills the entire plane. To measure the fractal dimension of a line (the classic example is an irregular, rocky coastline) we measure the length of the coastline at several different scales. We do this by "walking" a pair of dividers (or compass, i.e., a unit length ϵ) along a map of the coastline. If we then plot the logarithm of the total length against the logarithm of ϵ, we find that the plot defines a straight line (this was actually first done by L.F. Richardson in 1961). The line corresponds to the equation

$$L(\epsilon) = a\epsilon^{-D}$$

where D is a fractal dimension. D has a value between 1 and 2. The concept may be generalized to higher dimensions. For example, an irregular surface may have a fractal dimension between 2 and 3, and so on. Note that not every irregular line or surface has a (constant) dimension, but only those that display self-similarity. Though most fractals have non-integer dimensions, a fractal dimension may also be an integer. An example of a fractal with an integer dimension is the *Peano curve*, which (in the limit) completely fills a plane and so has dimension 2.

Since most strange attractors have fractal properties, numerical techniques have been developed to measure the fractal dimension of these attractors (e.g., Holden, 1986, Chapter 14). It has also been found that many strange attractors and other fractal objects (such as eddies in turbulent fluids) are not adequately described by a single dimension, but require a spectrum of dimensions (see **multifractal**). Other measures, for example, **lacunarity** describe different aspects of the geometry of fractal objects with the same fractal dimension.

True fractals are mathematical objects (**sets**) which can be fully explored only by using the concepts of pure mathematics, in fields such as set theory and topology. The properties of fractals can be considered to define a type of geometry, different from ordinary (Euclidean or Cartesian) geometry. Real objects resemble fractals only within a certain range (see **scale**); of course, it is equally true that real objects can be described by ordinary geometry only within a certain range (e.g., scales at which matter can be considered continuous and "smooth"), and to a certain degree of approximation.

Hamiltonian System

For a mechanical system, the Hamiltonian is a function of the coordinates and momenta of the parts of the system, and is equal to the total potential and kinetic energy of the system.

Systems are described as Hamiltonian systems if the Hamiltonian is a constant with time. In mechanics, frictionless (non-dissipative) systems are Hamiltonian. These are also systems in which the force fields may be described by potentials, that is, they may be expressed as the negative gradient of a scalar function (called the *potential*). In such a force field, the work done in moving a mass from one point to another does not depend on the path along which movement takes place.

Heaviside Function

A Heaviside Function, $H(t)$ is equal to zero if t is less than zero, and is equal to 1 if t is greater than or equal to zero.

Hénon Map (or Attractor)

A dynamical system described by the following two-dimensional mapping:

$$\begin{aligned} x_{n+1} &= 1 + y_n - Ax_n^2 \\ y_{n+1} &= Bx_n \end{aligned}$$

First described by Hénon (1976).

Hénon Quadratic Map

A Hamiltonian (area preserving) two-dimensional map described by:

$$\begin{aligned} x_{t+1} &= x_t \cos\alpha - y_t \sin\alpha + x_t^2 \sin\alpha \\ y_{t+1} &= x_t \sin\alpha - y_t \cos\alpha + x_t^2 \cos\alpha \end{aligned}$$

First described by Hénon (1969). A physical interpretation, as the return map of a periodically kicked rotor, has been given by Healy (1992).

Hénon and Heiles Map

One of the first area-preserving maps known to give rise to chaos. It is generated as a Poincaré section of the following Hamiltonian equations:

$$H = \frac{1}{2}\sum_{i=1}^{2}(p_i^2 + q_i^2) + q_1^2 q_2 - \frac{1}{3}q_2^3$$

$$\frac{\partial p}{\partial t} = -\frac{\partial H}{\partial q}, \quad \frac{\partial q}{\partial t} = \frac{\partial H}{\partial p}$$

First described, for an astrophysical model, by Hénon and Heiles (1964).

Intermittency

A signal is called intermittent "...if it is subject to infrequent variations of large amplitude." (Bergé et al., 1986, p.223). Intermittency is typical of transitions to turbulence in fluids, for example, velocity fluctuations measured close to the boundaries of turbulent boundary layers, jets or wakes.

Iterative (or Iterated) Map

A form of difference equation, where the next value $f(x_{n+1})$ of a function that depends on both x and t is calculated from the previous value $f(x_n)$. A classic example is the difference form of the **logistic equation** (see also **bifurcation**).

Jacobian

If we have a system of two differential equations such that

$$dx/dt = f(x,y)$$
$$dy/dt = g(x,y)$$

then the Jacobian is the matrix

$$J = \begin{pmatrix} \frac{\partial f}{\partial x} & \frac{\partial f}{\partial y} \\ \frac{\partial g}{\partial x} & \frac{\partial g}{\partial y} \end{pmatrix}$$

If (x,y) are the coordinates of a **fixed point** then the properties of the **eigenvalues** of J can be used to determine the stability of the fixed point. The concept is easily generalized to more than two dimensions. The basic idea is that the partial derivatives define the slopes towards or away from the fixed point in the immediate vicinity of that point in phase space, and that these slopes define the stability of the point.

KAM Theorem

This is a theorem in mathematics first postulated by Kolmogorov in 1954, and proved by Arnol'd (1963) and Moser (1962). It seems to have been first called the KAM theorem by Walker and Ford (1969: see that paper, or Lichtenberg and Lieberman, 1983, for original references). The basic idea is that Hamiltonian dynamical systems are not necessarily **ergodic**, that is, a trajectory does not necessarily "explore the entire region of phase space that is energetically available to it" (Lichtenberg and Lieberman, 1983, p.3). In an ergodic system, the trajectories are entirely chaotic. The theorem thus "set the stage" for further investigations of chaos in dynamic systems, because many real systems are neither entirely regular, nor entirely chaotic (**stochastic**). From a combination of theory and computation

> ...an extraordinary picture of the phase space for weakly perturbed systems has emerged. The invariant surfaces break their topology near resonances to form island chains. Within these islands the topology is again broken to form yet other chains of still finer islands. On an even finer scale, one sees islands within islands. But this structure is only part of the picture, for densely interwoven within these invariant structures are thin phase space layers in which the motion is stochastic ... For weakly perturbed systems with two degrees of freedom, the KAM surfaces isolate the thin layers of stochasticity from each other ... as the perturbation strength increases a transition can occur in which the isolating KAM surfaces disappear and the stochastic layers merge, resulting in globally stochastic motion that envelopes the phase space. Often the phase space divides into three regions. One region contains primarily stochastic trajectories. It is connected to a second, mainly stochastic region containing large embedded islands. A third region consists of primarily reg-

ular trajectories and is isolated from the first two by KAM surfaces (Lichtenberg and Lieberman, 1983, p.3–4).

Good examples of the phenomena described above are shown by the **Hénon map** and the **Hénon and Heiles map**.

Koch Snow Flake (or Curve)

This is one of the classic fractal curves, proposed by von Koch around 1904 (*fide* Voss, 1988). To construct a Koch curve an original straight line segment (for example, a side of an equilateral triangle) is replaced by a set of straight line segments, for example, the line segment is divided into three parts, and the central part is replaced by two lines equal in length to the original segment. The initial shape (the triangle in the example given) is called the *initiator* and the set of line segments that replace any original line segment, the *generator*. The process is carried out for all straight line segments in the initiator, and then repeated again for all the straight line segments in the transformed initiator, and so on indefinitely. If the process is carried through only a finite number of times, then the figure that results is not a true fractal, but a *prefractal*. In the equilateral triangle example, each straight line segment is replaced by four new segments ($N = 4$), each having one third of the original length ($r = 1/3$). The fractal (or similarity) dimension, D is then given by

$$N = 1/r^D$$

and

$$D = \log N / \log(1/r)$$

where N is the number of segments, and r is their length, relative to the original (i.e., N is the number of segments at the $(n+1)$th step of the generating process, divided by the number at the nth step). In the example

$$D = \log 4 / \log 3 = 1.26\ldots$$

Note that Koch curves are a whole class of curves (with different fractal dimensions) that may be generated by such algorithms. They are continuous, but not differentiable at any point. They are also self-similar in the sense that, at any stage of the generation process, each small segment reproduces exactly a section of the larger curve at some previous stage.

Legendre Transformation

This is a transformation between two forms of a partial differential equation. For example, in dynamics the momenta, p, may be expressed a function of the coordinates, q, the velocities, u, and time, t. To express u in terms of the other variables, we make use of a Legendre transformation (Percival and Richards, 1982). The transformations are also commonly used in thermodynamics, and they are used in obtaining the $f(\alpha)$ function used to characterize multifractals.

Lyapunov Functions and Coefficients

Suppose that the behavior of a system may be described by trajectories in **phase** (or state) **space**. Certain points or surfaces in the space may be stable equilibrium points or surfaces, that is, trajectories beginning near these points rapidly converge on them. Others may be characterised by unstable equilibrium, and so on. Certain functions, $V(\mathbf{x})$, where $\mathbf{x}$ is the vector of variables defining the phase space, may have the property that they describe smooth curves or surfaces along which the derivative of V with repect to time is always less than or equal to zero (remember that though all variables used to construct a phase space are functions of time, time is not one of the variables represented directly by the space). These functions are called *Lyapunov functions*. Lyapunov functions are useful for describing stability conditions in phase space.

For simplicity consider that the initial position in a **Poincaré section** through a phase space can be designated by a single variable, x. Then if the trajectory giving rise to this point by intersection with the Poincaré section is unstable, and we choose another trajectory separated from the first by a small distance Δx, we expect that, as the trajectories are traced through time (corresponding to successive intersections on the Poincaré section) the distance Δx will increase. If the increase is exponential we can represent Δx for the nth iteration as

$$\Delta x(n) = e^{n\lambda} \Delta x(0)$$

where λ is defined as the *Lyapunov exponent*. The size and sign of the exponent is therefore a measure of the instability of the trajectory. If the trajectory is a function of m state-space variables, there will be m Lyapunov exponents. Strange attractors, which show sensitivity to initial conditions, have at least one positive Liapunov exponent. A simple discussion is given in Holden (1986, Chapter 13).

Likelihood

This has a technical meaning in certain branches of statistics, (as in the *method of maximum likelihood*). The term is also used as a rough equivalent of probability.

Limit Cycle

In **phase** (or state) **space**, a limit cycle is any simple, closed trajectory that does not contain any singular points (i.e., there are no sigular points actually on the trajectory; there may be singular points inside the limit cycle trajectory). "Simple" implies that the trajectory cannot cross itself. Trajectories close to the cycle either outside or inside the limit cycle may converge on the limit cycle itself (which implies that the system is dissipative); if they converge on the cycle with increasing time then it is stable, if they diverge from it the cycle is unstable (Edelstein-Keshet, 1988, p.311–312). Note that a closed trajectory within a phase space is not a limit cycle unless there is local convergence towards or divergence from this trajectory. Stable limit cycles are a form of attractor.

Liouville's Theorem

This theorem states that for Hamiltonian systems, a small volume remains the same along trajectories in phase space. This expresses the "incompressibility of the flow" in phase space.

Logistic Equation

An equation describing growth of a some property x that is proportional to the product of x and the difference of x from some limit (e.g., 1)

$$dx/dt = rx(1-x)$$

It is a simple example of a nonlinear differential equation. The difference form of the logistic equation, called the *logistic map* is

$$x_{n+1} = rx_n(1-x_n)$$

and displays bifurcations for $r > 3.0$ and chaotic behaviour for some ranges of $r > 3.57$. See also **Feigenbaum number**.

Lorenz attractor (or equations)

See **attractor**. The equations are

$$
\begin{aligned}
dx/dt &= \sigma(y-x) \\
dy/dt &= rx - y - xz \\
dz/dt &= xy - bz
\end{aligned}
$$

The value of σ and b are generally set at 10 and 8/3 respectively. This system shows a variety of phenomena, depending on the parameter r. These include one steady state ($r < 1.0$), followed by a pitchfork **bifurcation** to two steady states ($r < 24.06$), and various degrees of chaos. The familiar "butterfly" attractor is generally shown for $r = 28$.

Manifold

This is a term from topology that defines classes of topological spaces: for most purposes a manifold may be considered to be a line, surface, or hypersurface. M is an n-manifold if for each point in M there is an open neighbourhood of the point, and a homeomorphism from the neighbourhood to an open set in the n-space. A *homeomorphism* is a special type of one-to-one **mapping** between two spaces.

Measure

In mathematics a measure is a real number associated with a set (and having certain other properties). Perhaps the most familiar example is probability which is associated with subsets forming part of an event space. Most introductory texts gloss over the fact that the ordinary methods of calculus are inadequate for probability theory. For example, ordinary (Riemann) integrals should be replaced by Lebesgue integrals, based on measure theory. Measure theory becomes even more important in dealing with fractal sets (see **multifractal**). For an introduction to measure theory, see Falconer (1990). A simple introduction to the Lebesgue integral is given by Stečkin (1964).

Mode Locking

See **phase locking, oscillator, circle map**.

Multifractal

Fractals are sets having the property of self-similarity: multifractals are measures associated with sets.

Multifractal measures are related to the study of a distribution of physical or other quantities on a geometric support. The support may be an ordinary plane, the surface of a sphere or a volume, or it could itself be a fractal. (Feder, p.66).

Multifractals arise when a measure defined on the support has different fractal dimensions on different parts of the support, in other words, where the measure shows spatial correlations. Such phenomena cannot be adequately described by a single fractal dimension, but require a complete spectrum of "dimensions". The concept was first developed by Mandelbrot for turbulence: however, he objects to using the term "dimension" for the "generalized dimensions" used in multifractal theory, and prefers to call them "scaling exponents."

The concept of multifractals has been applied to a wide range of mathematical and geophysical phenomena, including the properties of strange attractors and aggregation phenomena. The use of multifractals permits a much more complete description of phenomena than can be expressed by a single dimension, therefore their use has been increasing rapidly. For example, the structure of atmospheric turbulence, and the pore spaces in sandstones are known to show fractal geometry, and it is clear that they can only be described adequately by a model that includes spatial correlations. For an introduction and key references see Feder (1988) and Stanley and Meakin (1988).

A simple example of a multifractal is diffusion limited **aggregation** (DLA: Stanley and Meakin, 1988). In DLA, a "dendrite" grows by aggregation of particles onto a seed: the particles follow a random path, so that at a given stage of dendrite growth one can specify a probability, p_i, that the ith site on its surface will be hit by the next particle. The probabilities of all the sites constitute a probability distribution $n(p)$, which can be characterized by its moments, $Z(q)$ given by

$$Z(q) = \sum n(p).p^q$$

One can relate the moments to a characteristic diameter of the aggregate, L, by an exponential (or scaling) equation

$$Z(q) \sim L^{-\tau(q)}$$

It is generally assumed that only a small number of scaling exponents are necessary, in other words the τ reduce essentially to a few "dimensions" (which are probably fractal). It has now been shown, however, that this is not true of DLA, and an infinite number of exponents are required. Generally investigators represent these, not as the function $\tau(q)$ itself, but as its **Legendre transform** $f(\alpha)$. It has been found that this function is a very useful way to characterize multifractals. It is smooth and continuous and shows a single maximum. It is also universal, in that the same general type of function characterizes many different types of multifractal phenomena, or an identical function characterizes a whole range of phenomena. For example, a single $f(\alpha)$ distribution characterizes fluid turbulence formed in many different ways at many different scales (see Feder, 1988, p.76-78).

Another example is a multifractal constructed on a Cantor set. Instead of the usual (one scale) construction, we introduce two scales, e.g. $s_1 = 1/4$ and $s_2 = 2/5$. The unit line is first divided into a segment from 0 to 1/4, and a second segment from 3/5 to 1 (i.e., the first segment has length 1/4, the second has length 2/5). A probability of $p_1 = 3/5$ is assigned to the first segment, and of $p_2 = 2/5$ to the second segment (i.e., the first segment has a larger "weight" than the second). The process is then repeated, to give four segments with length $s_1^2, s_1 s_2, s_2 s_1$, and s_2^2, and with probabilities $p_1^2, p_1 p_2$, etc., and so on indefinitely. In this example, therefore, the support is a fractal set, the measure is a probability, and the spatial correlation is introduced by defining the probability in a way which depends upon position in the fractal set. The specified fractal set is then partitioned into subsets of length l_i, and to each subset it is also possible to assign a probability p_i. We assume that, in the limit as the subset length tends to zero, the probability is given by a power law

$$p = k\, l^\alpha$$

α is then a scaling exponent or (fractal) dimension defined for a particular item in the (fractal) set, and there is a function $f(\alpha)$ which shows how this dimension is distributed over the whole set. The process can be repeated for all items in the set, and the result is a summed distribution of $f(\alpha)$. Such a distribution can be calculated, or derived by computer simulation.

The term "multifractal" can be related either to the observation that a small number of dimensions is not adequate to describe multifractal phenomena or to the derivation of the measure (e.g., p in the Cantor example) from a union of many sets, each of which is fractal with its own dimension. Recently, **wavelets** have been used to calculate and illustrate the spatial properties of multifractals (Muzy et al., 1994).

Noise

It is useful to divide a time series into the signal (the "desired" or "meaningful" part) and the noise (the "undesired" or "meaningless" part). The distinction is subjective: one person's noise is another person's music (West and Shlesinger, 1990).

A common mathematical model for noise is given by:

$$y(t) = x(t) + \epsilon$$

where $y(t)$ is the total time series, $x(t)$ is the signal, and ϵ is the noise, assumed to be a **random variable**.

Certain types of noise may be defined precisely. Examples are *white noise*, which consists of a uniform random distribution, *Gaussian noise*, which is random (and so can be considered a variety of white noise) but has a Normal distribution.

Different types of noise may be distinguished using **spectral analysis**. The *variance (power) spectrum*, plots the variance $V(f)$ (also called the power or spectral density) against the frequency f, for a range of frequencies observed in the observed time series. White noise shows no systematic variation in variance with frequency. Much of the noise displayed by real spectra, however, shows that $V(f)$ varies as $1/f^n$.

If noise is generated by summing (integrating) white noise, as in a *random-walk*, then it is described as *Brownian*, or *Brown noise*. This noise shows a spectrum where $V(f)$ is proportional to $1/f^2$, that is, $n = 2$.

If we consider a time series composed only of random noise, then it is impossible to predict the next interval in the series from a knowledge of the series up to that time. It is still possible to state the probability that, in some time interval δt, the value of y will be less than some value y_o. Assume this probability is proportional to $\delta t H$, where H is called the *Hurst exponent*. For Brownian noise $H = (n-1)/2 = 0.5$.

A common type of noise, whose origin is not yet fully explained is $1/f$ noise, which shows a spectrum with $V(f)$ proportional to $1/f$, that is $n = 1$.

A significant feature of $1/f^n$ noise is that it implies that there is no scale that adequately characterizes the noise: all possible scales are present, though the larger scales are proportionately less frequent (see **scale, characteristic**).

The fractal dimension is given by

$$n = 5 - 2D \quad \text{or} \quad H = 2 - D$$

For example, Brownian noise is a fractal with dimension 1.5.

Suppose that a time series relates the magnitude of an event (e.g., a flood or an earthquake) to the time at which it occurs. Then, for a given length of record the frequency can be plotted against the magnitude: such plots often show a power law relationship, i.e. they plot as straight lines on log-log graph paper. Such observations imply a spectrum with $V(f)$ varying as $1/f^n$.

Oscillator

A system displaying oscillatory motion. The simplest oscillator is a mass whose motion is controlled by an ideal spring. Its equation of motion is given by

$$d^2x/dt^2 = -\omega^2 x$$

that is, the acceleration is proportional to the displacement (ω^2 is the spring constant). A general form of the equation of a nonlinear, damped oscillator is

$$d^2x/dt^2 = k_1 f_1(x) + k_2 f_2(x)$$

where the k's are coefficients, and the $f(x)$ are generally nonlinear functions. For a damped pendulum, for example, k_1 is $-g/L$, $f_1(x)$ is $\sin x$, k_2 is a (negative) damping coefficient, and $f_2(x)$ is dx/dt. Oscillators may be forced by adding another periodic driving force to the equation. An example is described by the **van der Pol equation**

$$d^2x/dt^2 = (\epsilon - x^2)dx/dt - \omega^2 x$$

The equation was derived by making the friction coefficient depend on the displacement, so that it was positive for large displacements but negative for small ones. Negative friction corresponds to forcing and is not meaningful for ordinary mechanical oscillators, but the equation does describe certain types of electrical oscillators. The equation has a **limit cycle**.

After an initial period of adjustment, driven oscillators may settle down to a regular oscillation where the periods of the two phases have been adjusted so that their ratio is rational: this is the phenomenon of **phase locking**.

Percolation Theory

The term was introduced by Broadbent and Hammersley (1957) for a new science dealing with clustering of particles by random processes. It includes the following problem: let particles occupy sites on a regular lattice (e.g., a two-dimensional square lattice), and define a cluster as a group of particles with nearest neighbour links to other particles in the cluster (that is, each particle connects above, below, or to the side with some other particle in the cluster). Assign a probability at random to each site, and then define whether a site is occupied or not by specifying a minimum probability, p, that any site is occupied by a particle. What must this probability level be so that there is a very high chance that a cluster spans the entire lattice (that is, it extends without a break from one side to another)? This can be defined as the percolation threshold for that lattice: if the lattice defines an idealized porous material, the "particles" are pore spaces, and the "clusters" are interconnected groups of pore spaces, then the percolation threshold would be the condition where a fluid could percolate through the porous material. It turns out that the transition from "non-percolating" to "percolating" lattices is generally quite abrupt (corresponding to a "critical" condition) and that, close to the critical condition, the sizes of clusters have interesting statistical properties: all possible sizes of cluster are present, and have the power law size distribution typical of fractals. Percolation theory applies not only to porous media but to aggregation of particles by random walk processes (thus to the formation of dendrites). It has much in common with the theory of phase transitions such as condensation from a vapour to a liquid or magnetization of minerals at the Curie Point (see **renormalization group**).

Phase Locking

...a resonant response occurring in systems of coupled oscillators or oscillators coupled to periodic external forces. In general, resonances occur whenever the frequency of a harmonic, $P\Omega_1$, of one oscillator approaches some harmonic, $Q\Omega_2$, of another; and in the resonant region the frequencies of the two oscillators lock exactly into the rational ratio P/Q (Jensen et al., 1984, p.1960).

See **circle map** and **oscillator**.

Phase Space

An ordinary differential equation of order n may be represented as a system of n first order ordinary differential equations. For example, the motion (in the x-direction) of a mass m suspended from a spring and subject to viscous damping (friction) can be described by the equation

$$d^2x/dt^2 + r(dx/dt) + \omega^2 x = -g$$

(see also **oscillator**). By letting $y = x$ and $z = dy/dt$ it is possible to write this equation as the system

$$
\begin{aligned}
dy/dt &= z \\
dz/dt &= -\omega^2 y - rz + g
\end{aligned}
$$

The set of all possible solutions of such a system is sometimes called a *flow* in the space of two dimensions. Its behaviour may be studied by plotting trajectories from some arbitrary starting point, using y and z as the coordinates of the space, rather than the original x and t. The variables y and z are called *state* variables, because they define the possible states of the system. In the example, y is the position, and z is the velocity. The space with these variables is called the *phase space*, and a section through it is called a **Poincaré section**. Strictly, a phase space is a plot of position vs. velocity (or momentum). The term is also used loosely for any space, defined by a set of state variables, related by first order differential equations. If the variables are not those of position and velocity, such a space is more properly called a **state space**.

Poincaré Section

The behaviour of a system of ordinary differential equations may be described by constructing trajectories in **phase (or state) space**. It is not always easy to describe the complete trajectory, and therefore it is better in some cases to study the intersection of the trajectories with a single plane.

Commonly the position of a point on the Poincaré section is determined numerically by a process of numerical integration, i.e, it is calculated from the position of the previous intersection by solving the differential equations for the trajectory. The process can be called a **mapping** (or iterative map) of the points and represented symbolically:

$$P_{k+1} = T(P_k)$$

Note, however, a difference between a Poincaré section and a typical iterative map. The map is generally calculated by incrementing time repeatedly by a constant amount. The time taken for trajectories to cross the same plane in phase space, however, is not necessarily the same for succesive trajectories.

One graphic representation of the chaotic nature of dynamic systems is the scatter of points, and the unpredictable way that they appear on Poincaré sections, or on iterative maps. Some of the general laws governing the distribution of chaotic and regular periodic trajectories in phase space are given by the **KAM theorem**.

Probability

Over the years, several different definitions have been suggested for the concept of probability. The most common approach at present is to define probability axiomatically, but the classical definition is also used.

The classical approach is to consider an experiment (or trial), such as tossing a coin, or throwing a die, which may have a limited number of outcomes (head, tail; 1,2,3,4,5,6). We may further define the outcomes as being equally likely (by inference, or more realistically, by definition). A particular outcome is called an event (see **random**). Then if a trial may result in any one of N exhaustive, mutually exclusive, and equally likely cases, and if M of these produce the event, then the probability is M/N. Probabilities vary from zero (impossible) to one (certain), and follow certain rules. Several problems with this classical approach will occur to the reader: it depends upon the prior definition of "equally likely outcomes", which is itself a concept involving probability, and there is the problem of how the concept can be applied if the number of outcomes is infinitely large (i.e., not enumerable).

The axiomatic theory is based on a set S of elementary events and a system B of subsets of S. The elements of B (subsets of S) are called random events, and B is called a *Borel field* if it satisfies the following conditions:

1. S is an element of B;

2. If two sets $E1$ and $E2$ are elements of B then their union, their intersection, and their complements are also elements of B;

3. If the sets $E1, E2, \ldots$ are elements of B then their union and their intersection are also elements of B.

Then Kolmogorov's system of axioms is:

1. To every random event in the field of events there is assigned a non-negative real number called the probability;

2. The probability of the certain event is 1;

3. If the events $E1, E2, \ldots$ are mutually exclusive in pairs, then the probability of their intersection is equal to the sum of their individual probabilities.

Random

This simple term proves to be a hard one to define. If we conduct an experiment, which may be something as simple as flipping a coin, or drawing a ball from an urn containing a mix of black and white balls, and if we observe that the result of the experiment is not completely reproducible, and not completely predictable from a knowledge of previous results of running the same experiment, then we may say that the result is at least partly produced by random events.

If it is possible to enumerate all possible outcomes of a random experiment, then this set can be called the *sample space*. For example, in tossing a coin we recognize heads (H) and tails (T) as the two possible outcomes. Then an *event* is a subset of a sample space. For example, if we toss the coin once, the sample space consists of one head and one tail, and the event will be one or the other (this is an *elementary event*). If we toss the coin twice in succession the sample space would be: HH, HT, TT, TH. An event would then be one of these four possibilities. Instead of using words or letters we could assign a number to each outcome. Then we could define a *random variable* as a function connecting a set of outcomes to the experiment to the set of numbers that correspond to these outcomes. A little more formally we say that a random variable $f(z)$ describable by a sample point, z, is a function defined on the sample space of an experiment such that for every real number of the variate, z, there exists a probability $p(z)$.

We see that the concept "random" is closely connected with the concept of sampling. Sampling implies the (at least potential) existence of a set of possible events, which is the object of investigation, and the actual subset of observed events. The former is the *population*, and the latter is the *sample*.

A sample may be defined as a *random sample* if every item in the population has an equal chance of being chosen. This further implies that the result of one "choice" (trial, draw, experiment) does not influence in any way the result of the next one. Alternately, a random sample may be defined as a set of independently and identically distributed random variables.

An interesting critique of the distinction generally drawn between "random" or "stochastic" processes, and deterministic processes has been given by Ford (1983). One problem is that some fully deterministic processes give rise to outcomes which, to an observer who does not know the generating process, appear to be completely unpredictable and therefore "random". In fact, since a coin toss follows simple dynamic laws, we might say that it is exactly such a process. Interestingly enough, the pioneers of statistics actually conducted long coin-toss experiments, which "verified" experimentally some of the fundamental theorems of probability theory (e.g., the law of large numbers). However, such theorems can no more be verified by experiment than the theorems of geometry can be. All that can be verified is that, under certain experimental conditions, the axioms of probability theory, and deductions from them, correspond reasonably closely to observable phenomena.

Renormalization Group

A scaling technique developed originally in quantum theory, and applied by K.G. Wilson to explain phase transitions (see Wilson, 1979). "Renormalization" means the repeated rescaling of variables so that the general form of the equations connecting them is preserved. The variables constitute a "group" in the mathematical sense of a set whose elements can be combined together (in this case, rescaled) in a way having definite mathematical properties. Application of the technique reveals that close to phase transitions, physical variables can be related by power equations, and that there are certain values for the exponents in these equations that remain invariant under repeated rescaling: these values define the **fixed points** of the system, and may have very general significance, which is independant of the details of the theoretical model used to derive them.

Renormalization-group methods have been applied to percolation phenomena, many of which have fractal properties. The self-similarity of fractals makes them a natural field of application of these methods. They have also been applied to the sequence of bifurcations shown by the **logistic map**. Simple introductions are given by Wilson (1979), Feder (1988), Gould and Tobochnik (1988, v.2), Stauffer (1985), Strogatz (1994), Turcotte (1992) and Zallen (1983). For a historical summary see Nelson (1981).

Return Map

Given a map $x_{k+1} = f(x_k)$, a first return map plots values of x_k against x_{k+1}. It is also possible to plot return maps for higher iterations (e.g., k and $k+2$). Complicated strange attractors may be easier to visualize by displaying two dimensional sections through them. But an even more striking simplicity is often seen by plotting the first return map. As this simplicity is not expected from a truly random map, its existence is evidence for nonlinear dynamical chaos. The simplicity of the return map may also allow it to be characterized by a numerically fitted function, say g. Then it is possible to determine the largest **Lyapunov exponent** of the original system from this function. See Bergé et al. (1986, Appendix B) for details.

Rössler attractor (or equations)

A particularly simple set of equations with a strange attractor, first proposed in 1976. The equations are

$$
\begin{aligned}
dx/dt &= -(y+z) \\
dy/dt &= x + ay \\
dz/dt &= b + z(x-c)
\end{aligned}
$$

Scale, Characteristic

Individual phenomena generally may be partly described by a scale: for example, a rock fragment may be characterized by its volume or diameter, and an earthquake by its Richter magnitude. An observed sample of such phenomena shows a range of such scales, but the variation can be described by statistical measures such as the mean and standard deviation. It is often assumed that the entire natural population also can be characterized by such scales: this would be correct, for example if the size of rock fragments was Normally (or Lognormally) distributed. But if the size follows a power law distribution, that is, if it shows a fractal distribution, then there is no characteristic physical scale. There is no meaningful "average size" of a true fractal size population.

The notion that there is no characteristic scale to a group of phenomena is counterintuitive, because our experience indicates that even if a part of the distribution is fractal, in nature there must be upper and lower limits. For, example rock fragments cannot be smaller than the unit cell of mineral grains making up the rock, or larger than the entire earth. But observation indicates that many phenomena are fractal over a very large range.

The fact that fractal distributions have no characteristic scale suggested to Bak et al. (1987) that they indicate a system maintained in a critical state. It is known that critical transitions, such as phase transitions, the Curie point in magnetism, or the critical state in percolation theory, are characterized by the presence of all possible scales of phenomena (e.g., all possible sizes of ordered domains or interconnected clusters of pore spaces). Bak has suggested, therefore, that if all possible scales are continuously present, the system is probably in a critical state, and must be maintained in this state by the operation of the system itself. In this sense, the system shows evidence of *self-organized criticality*.

Self-Affinity

When a geometric object, for example, a cube is scaled to a different size, each of its dimensions is scaled by the same factor. Such a scaling can be described as due to a similarity transformation. Objects which are strictly similar to each other can be transformed into each other by such a transformation (see **self-similarity**).

Under an affine transformation, different dimensions are scaled by different factors. If a part of an object can be transformed into a smaller (or larger) part of the same object using an affine transformation the object is described as described as self-affine. Just as with similarity, affinity may be "absolute" or it may be statistical. In statistical affinity the parameters describing the two statistical distributions undergo an affine transformation.

Self-Organization

In the context of chemical systems, this term was popularized by Nicolis and Prigogine (1977). Strangely, though the term appears in the title of their books, it is nowhere given a succinct definition. Perhaps their closest approach is the statement (p.5):

> . . . self-organization in nonequilibrium systems [is] characterized by the appearance of dissipative structures

through the amplification of appropriate fluctuations.

Paradigms include purely physical systems, such as thermal convection in fluids, and oscillating chemical systems such as the Belousov-Zhabotinski reaction, and Liesegang phenomena.

Ortoveva et al. (1987) have given a general discussion of self-organization in geochemical reaction-transport systems, and give the following definition (p.980):

> Self-organization is the autonomous passage of a system from an unpatterned to a patterned state without the intervention of an external template.

They point out two necessary conditions: that the system be far from equilibrium (therefore controled by nonlinear processes) and that "at least two processes active in the system be coupled" (p.980).

The study of self-organization therefore overlaps to a large extent with the study of nonlinear dynamics, though it concentrates on the appearence of pattern, rather than on chaotic phenomena. See also **Brusselator**.

Self-Similarity

This is a key concept in fractals. Self-similarity means that a part of the whole shows the essential features of the whole. In the case of fractal curves, such as the Koch Curve the similarity may be exact, in the sense that a small part of the curve may be magnified to reproduce exactly the features of a larger part of the curve. The self-similarity of many natural fractals, however, is statistical rather than exact, that is, small parts of the system shows the same statistical properties as larger parts. This implies that the system has no characteristic physical scale (see also self-affinity).

...the concept of self-similarity is certainly not new ...It has been well known since the detailed study of critical properties of phase transitions ...and has been instrumental in the formulation of the renomalization group approach ...that has essentially solved this problem. In this case, however, self-similarity was considered a peculiarity of the competition between order and disorder at a particular temperature for equilibrium phase transitions. We can now see that this property is much more common and appears in many equilibrium and nonequilibrium phenomena apparently unrelated to the problem of phase transitions. (Pietronero, 1988, p.278)

Sierpinski Gasket and Carpet

The Sierpinski gasket is a fractal produced by starting with a filled triangle, cutting out a central triangle (with a side one half of the original), and then repeating the process indefinitely. The fractal dimension is 1.58...The carpet is similar except that one begins with a square, and cuts out a central square with a side one third of the original. The fractal dimension is 1.89...These and other Sierpinski figures can also be generated in other ways.

Spectral Analysis

Spectral analysis is a way of representing a time series $x(t)$ as a combination of a set of sinusoidal waves of different frequencies f and amplitudes a. Such an infinite series of sinusoidal terms is called a *Fourier series*. The (*power* or *variance*) spectrum shows the contribution of each frequency (which is proportional to a^2 to the total variance of x. The phases (i.e., starting times) of the sinusoidal components are ignored:

each component has its own frequency (or wavelength) and amplitude, but these do not vary with time.

This form of representation and analysis of time series has many advantages, expounded in texts on the subject (e.g., Newland, 1993). For example, spectral anaysis has been used to reveal the main periodicities present in climatic time series (the *Milankovich cycles*), and can be used to "smooth" or "filter" time series, and to determine the **fractal dimension** of an time series, such as a **random walk** or a topographic profile (see **self-affine fractal**).

Spectral analysis, is however, a linear technique of analysis (each term in the series is orthogonal to the others, so the whole series does not have to be recomputed when any one term is added or dropped). As a consequence, it is not a good technique for analysing time series generated by nonlinear dynamical systems. The spectra of time series produced by nonlinear chaotic systems are similar to those of broadband **noise**.

Spectral analysis analyses periodicities in terms of symmetrical, sinusoidal waves, so it is not very useful to analyse trains of strongly asymmetrical waves, such as bedforms. Spectral analysis assumes that the waveform does not change in wavelength or amplitude along the time series—which is not true of many natural quasi-periodic phenomena. A technique that potentially avoids some of the pitfalls is **wavelet analysis**.

State Space

See **phase space**.

Stochastic

The word is derived from the Greek, meaning to contemplate or conjecture, and its use in mathematics can be traced back to Jacob Bernoulli. The term is often used as a synonym of "random", but generally as applied to a "process" (e.g., stochastic process, random process). Such a process is generated by one or more random variables (see **random**) which in turn depend on certain space and/or time parameters. For example, a process that gives rise to a "random walk" is a stochastic process. The outcome of a stochastic process, therefore, cannot be exactly predicted: at best, it is possible to associate probabilities with the range of possible outcomes. Most stochastic processes attempt to combine purely deterministic phenomena with a degree of random behaviour. For example, a stochastic theory of ballistics might combine the usual deterministic theory based on Newtonian mechanics, with estimates of the probable error in prediction of such a theory based upon the addition to the theory of certain random variables.

Symbols

Much of the literature of Nonlinear Dynamics makes use of concepts and symbols of branches of pure mathematics that are not well-known to many scientists. Useful reference works that explain these concepts include Boroski and Borwein (1989), Daintith and Nelson (1989) and Jackson (1989).

The following is a short list of the most commonly used notations.

- **Sets:** A set consists of a collection of defined elements, e.g., the set of real numbers $\mathcal{R}$, or a set of points in a two-dimensional Euclidian space $\mathcal{R}^\in$. The set of elements x, having a defined property $P(x)$ may be written

$$\{x|P(x)\}$$

For example, if x lies in the range $0 < x < 1$, we write

$$\{x|0 < x < 1\}$$

That x is an element of the set X is written $x \in X$. $X \subset Y$ means that set X is a subset

of set Y. $X \cup Y$ is the *union* of sets X and Y, i.e. the set consisting of all the elements of both X and Y. $X \cap Y$ is the *intersection* of sets X and Y, i.e. the set consisting of all the elements that are common to both X and Y. The *Cartesian product* of two sets X and Y is written $X \times Y$ and consists of the set of all ordered pairs (x, y). This definition can be written

$$X \times Y = \{(x, y) : (x \in X)\&(y \in Y)\}$$

- **Functions:** The familiar notation

$$f(x) = \mu x(1 - x)$$

defines a function, f of x. We generally think of functions as continuous, in the sense that they can be represented (in two dimensions) by a smooth line. A continuous function is abbreviated C^n, where n refers to the highest order derivative that the function possessess at every point in the domain of interest.

- **Maps:** The concept of a function is frequently extended to discrete maps, such as

$$x_{k+1} = \mu x_k(1 - x_k)$$

which defines the value of x for the $(k+1)$th step, in terms of its value at the kth step. A map associates each element x of one set X with an element y in a second set Y. This can be written

$$f : X \rightarrow Y \quad \text{or} \quad f|X \rightarrow Y$$

or

$$f : x \mapsto y$$

In words, $x \mapsto y$ means "x maps to y".

- **Composition:** Repeated application of a discrete mapping can be represented

$$f^n(x) = f \circ f \circ \ldots f$$

Note that this does *not* mean f raised to the nth power, but f applied n times to some starting value of x.

Torus

A doughnut-shaped surface. The ideal torus is formed by revolving a circle about an axis which is in the plane of the circle but does not intersect the circle. The term may be used by extension to any surface in phase space, whose Poincaré section lies on a closed loop. Trajectories in the three-dimensional phase space wind around the surface of the torus, which constitutes a **limit cycle**. Such trajectories are characterized by two frequencies, one related to motion around the torus itself (as seen in a "plan view" or axial projection) and the other to motion around sections through the torus. The *winding number* is the ratio of these two frequencies: if it is rational, then the trajectories repeat themselves after a number of paths around the torus and the Poincaré section consists of a number of discrete points; if it is irrational, then the trajectories cover the entire surface of the torus, and the Poincaré section consists of a continuous line that forms a loop.

Van de Pol equation

A second order differential equation with a limit cycle. See **oscillator**.

Wavelets

Fourier series analyse a time series into a sum of sinusoidal wave components. The amplitude of the components may vary, but the frequency or wavelength do not vary with time. Thus Fourier or **spectral analysis** can only be applied to *stationary* time series, i.e., those whose statistical properties do not vary with time.

Wavelets are waveforms that are localized in time, at least to the extent that their amplitudes decrease asymptotically to zero away from some time origin. They are also required to have certain other mathematical properties (e.g., the integral of the entire wavelet is zero). The number of different wavelets can be defined is very large, in fact infinite. A time series may be analysed as a sum of a given form of wavelet that has been translated along the series, and dilated (changed in scale): thus, wavelets are suitable for the analysis of non-stationary time series. Meyer (1993) distinguishes between "time-scale" and "time-frequency" analysis. In time-scale analysis, the wavelet is allowed to change wavelength as it is translated. In time-frequency analysis, the frequency changes as the wave is translated.

The practical use of wavelets originated in geophysics, because a seismic signal originates as a finite waveform: in exploration seismology, the form of the original wavelet is known and can even be controlled by the seismologist. So it made sense to try to analyse the returning signals in terms of this waveform. In the last ten years, wavelets have been intensely studied by mathematicians, and applied to signal processing, image analysis and compression, and the description of fractals, multifractals and turbulence (Muzy et al., 1994).

Acknowledgments

An earlier draft was improved by comments from David Goodings.

References

ARNOL'D, V.I., 1986, Catastrophe Theory, second Edition: New York, Springer-Verlag, 108 p.

BAK, Per, 1986, The devil's staircase: Physics Today, December 1986, p.38–45.

BAK, Per, TANG, C., and WIESENFELD, K., 1987, Self-organized criticality: an explanation of 1/f noise: Physical Review Letters, v.59, p.381–384.

BERGÉ, P., POMEAU, Y., and VIDAL, C., 1986, Order within Chaos: New York, John Wiley, 329 p.

BOROSKI, E.J., and BORWEIN, J.M., 1989, Dictionary of Mathematics. London, Collins, 659 p.

BROADBENT, S.R., and HAMMERSLEY, J.M., 1957, Percolation processes I. Crystals and mazes: Proc. Cambridge Philosophical Soc., v.53, p.629–641.

CRUTCHFIELD, J.P., FARMER, J.D., PACKARD, N.H., and SHAW, R.S., 1986, Chaos: Scientific American, v.255, no. 6, p.46–57.

DAINTITH, J., and NELSON, R.D., eds., 1989, The Penguin Dictionary of Mathematics: London, Penguin Books, 350 p.

ESSEX, C., and NERENBERG, M.A.H., 1990, Fractal dimension: limit capacity or Hausdorff dimension? Amer. Jour. Physics, v.58, p.986–988.

FALCONER, K., 1990, Fractal Geometry: Mathematical Foundations and Applications: New York, John Wiley and Sons, 288 p.

FEDER, J., 1988, Fractals: New York, Plenum Press, 283 p.

FORD, J., 1983, How random is a coin toss? Physics Today, April 1983, p.40–47.

GLASS, L., and MACKEY, M.C., 1988, From Clocks to Chaos: The Rythms of Life: Princeton Univ. Press, 248 p.

GOULD, H., and TOBOCHNIK, J., 1988, An Introduction to Computer simulation Methods: Applications to Physical Systems: Reading, MA, Addison-Wesley Publ. Co., 2 volumes.

HASTINGS, S.P., 1994, A "shooting" approach to chaos: Internatl. Jour. Bifurcations and Chaos, v.4, p.17–32.

HEALY, J.F., 1992, A physical interpretation of the Hénon map. Physica D, v.57, p.436–446.

HÉNON, M., 1969, Numerical study of quadratic area-preserving mappings: Quarterly of Applied Math., v.27, p.291–311.

HÉNON, M., 1976, A two-dimensional mapping with a strange attractor: Communications in Mathematical Physics, v.50, p.69–77.

HÉNON, M., and HEILES, C., 1964, The applicability of the third integral of motion: some numerical experiments: Astronomical Jour., v.69, p.73–79.

HOLDEN, A.V., ed., 1986, Chaos: Princeton Univ. Press, 324 p.

JACKSON, E.A., 1989, Perspectives of Nonlinear Dynamics, Volume 1: Cambridge, Cambridge University Press, 495 p. (For symbols, see Appendix A and B, p.376–392).

JENSEN, M.G., BAK, P., and BOHR, T., 1984, Transition to chaos by interaction of resonances in dissipative systems. I. Circle maps: Physical Review, v.A30, p.1960–1969.

LI, T.Y., and YORKE, J.A., 1975, Period three implies chaos: Amer. Mathematical Monthly, v.82, p.985–992.

LICHTENBERG, A.J., and LIEBERMAN, M.A., 1983, Regular and Stochastic Motion: New York, Springer-Verlag, 499 p.

MANDELBROT, B.B., 1982, The Fractal Geometry of Nature: New York, W.H. Freeman and Co., 468 p.

MEYER, Y., 1993, Wavelets: Algorithms and Applications: Philadelphia, SIAM, 133 p.

MUZY, J.F., BACRY, E., and ARNEODO, A., 1994, The multifractal formalism revisited with wavelets: Internatl. Jour. of Bifurcation and Chaos, v.4, no.2, p.245–302.

NELSON, D.R., 1981, Renormalization: in R.G. LERNER and G.L. TRIGG, eds., Encyclopedia of Physics: Reading, MA, Addison-Wesley Publ. Co., p.876–878.

NEWLAND, D.E., 1993, An Introduction to Random Vibrations, Spectral and Wavelet Analysis, third Edition: New York, John Wiley, 477 p.

NICOLIS, G., and PRIGOGINE, I., 1977, Self-Organization in Nonequilibrium Systems. From Dissipative Systems to Order through Fluctuations: New York, John Wiley and Sons, 491 p.

ORTOLEVA, P., MERINO E., MOORE, C., and CHADAM, J., 1987, Geochemical self-organization I: reaction-transport feedbacks and modeling approach: American Jour. Sci., v.287, p.979–1007.

PERCIVAL, I. and RICHARDS, D., 1982, Introduction to Dynamics: Cambridge, Cambridge Uniiversity Press, 228 p.

PIETRONERO, L., 1988, Fractals in physics: introductory concepts: in S. Lundqvist, N.H. March and M.P. Tosi, eds., Order and Chaos in Nonlinear Physical Systems. New York, Plenum Press, p.277–294.

PRIGOGINE, I., 1980, From Being to Becoming: Time and Complexity in the Physical Sciences: New York, W.H. Freeman and Co., 272 p.

RUELLE, D., and TAKENS, F., 1971, On the nature of turbulence: Communications in Math. Physics, v.21, p.167–192.

RUELLE, D., 1980, Strange attractors: Math. Intelligencer, v.2, p.126–137.

SANDER, L.M., 1987, Fractal growth: Scientific American, v.256, no. 1, p.94–100.

SAUER, T., YORKE, J.A., and CASDAGLI, M., 1991, Embedology: Jour. Statistical Physics, v.65, p.479–616.

SMALE, S., 1967, Differentiable dynamical systems: Bull. Amer. Math, Soc., v.13, p.747–817.

SORENSEN, P.R., 1984, Fractals: Byte, Sept. 1984, p.157–172.

STANLEY, H.E., and MEAKIN, P., 1988, Multifractal phenomena in physics and chemistry: Nature, v.335, p.405–409.

STAUFFER, D., 1985, Introduction to Percolation Theory: Philadelphia, Taylor and Francis, 124 p.

STEČKIN, S.B., 1964, Theory of functions of a real variable: Chapter XV in A.D. Aleksandrov, A.N. Kolmogoroff, and M.A. Lavrent'ev, eds., Mathematics: Its Content, Methods, and Meaning. New York, Dorset Press, v.3, p.3–36.

STROGATZ, S.H., 1994, Nonlinear Dynamics and Chaos, with Applications to Physics, Biology, Chemistry and Engineering: Reading, MA, Addison-Wesley Publ. Co., 498 p.

THOMPSON, J.M.T., and STEWART, H.B., 1986, Nonlinear Dynamics and Chaos: New York, John Wiley and Sons, 376 p.

THOMPSON, J.M.T., and STEWART, H.B., 1993, A tutorial glossary of geometrical dynamics: Internatl. Jour. Bifurcations Chaos, v.3, p.223–239.

TURCOTTE, D.L., 1992, Fractals and Chaos in Geology and Geophysics: Cambridge Univ. Press, 221 p.

WALKER, G.H., and FORD, J., 1969, Amplitude instability and ergodic behavior for conservative nonlinear oscillator systems: Physical Review, v.188, p.416–432.

WEST, B.J., and SHLESINGER, M., 1990, The noise in natural phenomena: American Scientist, v.78, p.40–45.

WILSON, K.G., 1979, Problems in physics with many scales of length: Scientific American, v.241, no.2, p.158–179.

ZALLEN, R., 1983, The Physics of Amorphous Solids: New York, Wiley-Interscience, 304 p.

ZEEMAN, E.C., 1976, Catastrophe theory: Scientific American, v.234, no.4 (April), p.65–83.

A useful introduction to many of the basic ideas (particularly those of phase spaces and iterative maps) is given by:

EDELSTEIN-KESHET, Leah, 1988, Mathematical Models in Biology: New York, Random House, 586 p.

Some of the statistical definitions are after:

TIETJEN, G.L., 1986, A Topical Dictionary of Statistics. New York, Chapman and Hall, 171 p.

Appendix II: Chaos and Fractal Software

Gerard V. Middleton
Department of Geology
McMaster University
Hamilton ON L8S 4M1, Canada

The following is a list of sources of programs. The programs may be available on disk (with or without source code), or as computer programs in various languages, or as algorithms clearly written out in pseudocode. Most of these are demonstration programs: a few are for analysing data time series, and these have been marked with an asterisk. Prices are not given, but range up to $100.

I make this list available without any guarantees or recommendations. I have not seen or implemented many of these programs. Note that reviews of mathematical software appear regularly in the *Notices of the American Mathematical Society*, and in *Computers in Physics*.

This list includes only a few examples of software available on the Internet from ftp sites, or by request from the authors. Sites known to offer such software are:

`spanky.triumf.ca`

`inls.ucsd.edu`

`ftp.santafe.edu`

See also the FAQ (Frequently Asked Questions) for the Usenet discussion group Sci.nonlinear.

ARONSON, J.W., 1990. CHAOS: a SUN-based program for analysing chaotic systems. Computers in Physics, July/August 1990, p.408–417.

BACK, A., GUCKENHEIMER, J., MYERS, M., WICKLIN, F., and WORFOLK, P., 1992, dstool: Computer assisted exploration of dynamical systems: Notices of the American Math. Soc., v.39, no.4, p.303-309. (Describes a SUN Unix workstation program, using the X11 system, for the investigation of dynamical systems. Full documentation and source code are available by ftp from `macomb.cam.cornell.edu`, or by mail from the Center for Applied Mathematics, Cornell University, Ithaca NY 14853. Version 1.1 is now available).

BAKER, G.L., and GOLLUB, J.P., 1990, Chaotic Dynamics: an Introduction: Cambridge University Press, 182 p. (Includes some simple BASIC programs).

BARNSLEY, M.F., 1989, The Desktop Fractal Design System: New York, Academic Press, 38 p. + diskette (`.exe`, a few short source codes in Turbo BASIC).

BECKER, K.-H., and DÖRFLER, J., 1989, Dynamical systems and fractals: Computer Experiments in Pascal: Cambridge University Press, 398 p. (Demonstrations, source code included).

BESSOIR, T., and WOLF, A., 1990, Chaos Simulations: Physics Academic Software (90 p. manual + diskette for IBM PCs.

160

Demonstrates logistic map, Lyapunov exponents, billiards in a stadium, sensitive dependence, n-body gravitational motion. No source code.)

*ELLNER, S., NYCHKA, D.W., and GALLANT, A.R., 1992, LENNS, a program to estimate the dominant Lyapunov exponent of noisy nonlinear systems from time series data: Raleigh NC, Carolina State University, Statistics Mimeo Series # 2235 (BMA Series # 39).

ENLOE, C.L., 1989, Solving coupled, nonlinear differential equations with commercial spreadsheets: Computers in Physics, Jan/Feb 1989, p.75–76.

*FOWLER, A.D., and ROACH, D.E., 1993, Dimensionality analysis of time-series data: nonlinear methods: Computers and Geoscience, v.19, p.41–52 (includes source code in Pascal for Macintosh).

GOULD, H., and TOBOCHNIK, J., 1988, An Introduction to Computer Simulation Methods. Applications to Physical Systems. Part 1 and Part 2: Reading, MA, Addison- Wesley Publ. Co., 695 p. (Includes code in True BASIC)

GUCKENHEIMER, J. and KIM, S., 1990, KAOS: Cornell University, Mathematical Sciences Institute Tech. Rept. (Program for Sun workstations—now replaced by dstool: see BACK et al., 1992)

HAROLD, J.B., 1991, Chaotic Mapper: Physics Academic Software, 108 p. manual + diskette for IBM PC. (Demonstrates 22 1D–2D maps and 3D differential equations. Equations can also be added. Fractal systems include the Mandelbrot and Julia sets. No source code.)

*HASTINGS, H.M., and SUGIHARA, G., 1993, Fractals: a User's Guide for the Natural Sciences: Oxford University Press, 320 p. (Includes source code in Turbo Pascal.)

HUGHES, G., 1986, Hénon mapping with Pascal: Byte, v. 11, no. 13 (December 1986), p.161–178.

*JAGGI, S., QUATTROCHI, D.A., and LAM, N.S.-N., 1993, Implementation and operation of three fractal measurement algorithms for analysis of remote-sensing data: Computers and Geosciences, v.19, p.745–767.

KOÇAK, H., 1989, Differential and Difference Equations through Computer Experiments: with a supplementary diskette containing PHASER: An Animator/Simulator for Dynamical Systems for IBM Personal Computers: New York, Springer-Verlag, 224 p. (Demonstrates a large number of 1D–4D differential equations—many not chaotic—and 1D–3D difference equations. It is also possible to add equations not in the existing library. No source code, no programs for data analysis.)

MEISS, J.D., Std Map. (For Macintosh. Iterates 7 different area preserving maps: finds periodic orbits, cantori, stable and unstable manifolds, and allows you to iterate curves. Available by anonymous ftp from `ftp://amath.colorado.edu/pub/dynamics/programs/`)

MIDDLETON, G.V., ed., 1991, Nonlinear Dynamics, Chaos and Fractals, with Applications to Geological Systems: Geological Association of Canada, Short Course Notes v.9, 235 p. + diskette (with source and exe. code in Turbo Pascal. Includes an implementation of the Grassburger-Procaccia algorithm.)

NUSSE, H.E., and YORKE, J.E., 1994, Dynamics: Numerical Explorations: Springer-Verlag, 484 p. + diskette (can be used for demonstration or to study dynamical systems: systems not in library can be added. No programs for data analysis.)

PANKRATOV, Kirill, 1995, MatLab Chaos: (A collection of routines for correlation dimension calculations. Requires MatLab. Available by anonymous ftp from `http://puddle.mit.edu/ kirill/PUB/ MATLAB/CHAOS`)

PARKER, T.S., and CHUA, L.O., 1990, Practical Numerical Algorithms for Chaotic Systems: New York, Springer-Verlag, 348 p. (pseudocode)

PEITGEN, H.-O., and SAUPE, D., eds., 1988, The Science of Fractal Images: New York, Springer-Verlag, 312 p. (has a few algorithms in pseudocode).

PEITGEN, H.-O., JÜRGENS, H., and SAUPE, D., 1992, Chaos and Fractals: New Frontiers of Science: New York, Springer-Verlag, 984 p. (some programs in BASIC)

ROLLINS, R., WEEKLEY, D., and De SOUSA-MACHADO, S., 1991, Chaotic Dynamics Workbench: Physics Academic Software (78 p. manual + diskette for IBM PC. Demonstrates dynamical systems including Duffing two-well, soft-spring, hard-spring, and Ueda oscillators; driven, damped plane pendulum; Lorenz and Hénon-Heiles systems.)

*RUSS, J.C., 1994, Fractal Surfaces: New York, Plenum Press, 309 p. + diskette (`.exe` and source code for both PC – source in C, and Mac – source in Pascal).

SPROTT, J.C., 1993, Automatic generation of strange attractors: Computers and Graphics, v.17, p.325-332 (Source code in Quick-BASIC)

SPROTT, J.C., 1993, Strange Attractors: Creating Patterns in Chaos: New York, M&T Books, 426 p. + diskette. (Programs in BASIC and C for demonstrating attractors, many not available in other demonstrations.)

SPROTT, J.C., and ROWLANDS, G., 1990, Chaos Demonstrations: Intellimation Library for the Macintosh.

SPROTT, J.C., and ROWLANDS, G., 1990, Chaos Demonstrations 2.0: Physics Academic Software, 94 p. manual + diskette for IBM PCs. (Demonstrates 22 systems, including fractals, cellular automata, anaglyphs, the driven pendulum, nonlinear and Duffing oscillators, and the magnetic quadrupole. No source code.)

*SPROTT, J.C., and ROLLINS, G., 1992, Chaos Data Analyser: Physics Academic Software, 52 p. manual + diskette for IBM PC. (Software for analysing time series by calculating summary statistics, plotting histograms, fitting polynomials, power spectral and dominant frequency analysis, autocorrelation, embedding, state space display, return maps, singular value decomposition of the correlation matrix, and calculating Lyapunov exponents, capacity dimension, correlation dimension using the Grassberger-Proccacia method.)

STEVENS, R.T., 1990, Fractal Programming in Turbo Pascal: Redwood City, CA, M&T Books, 462 p.

TUFARILLO, N.B., ABBOTT, T., and REILLY, J., 1992, An Experimental Approach to Nonlinear Dynamics and Chaos: Addison-Wesley, 340 p. + diskette. (Two programs for Macintosh including source code in C.)

162

Adresses (other than regular publishers)

Physics Academic Software
c/o The Academic Software Library
Department of Physics
Box 8202
North Carolina State University
Raleigh NC 27695-8202
Tel: (800) 955-TASL FAX: (919) 515-2682

Geological Association of Canada
Memorial University of Newfoundland
St. Johns NF A1B 3X5
Canada

Appendix III: Computer Programs in Diskette

Gerard V. Middleton
Department of Geology
McMaster University
Hamilton ON L8S 4M1, Canada

INTRODUCTION

Inside the back cover of this volume there is a diskette, formatted for MS-DOS, and suitable for running in most IBM-compatible computers. It contains a number of programs written by Gerry Middleton in Turbo Pascal (Copyright Borland International, 1984, 1989). Certain support files are also included, notably the `.bgi` graphics files, which remain the property of Borland International and may be be sold or distributed separately. The `.bgi` files must be in the default directory for the programs to run correctly—the simplest way to achieve this is to run the programs directly from the diskette, making the drive containing the diskette the default drive.

The programs run best on computers equipped with VGA color graphics. A math coprocessor is not necessary: many of the programs may also run on machines with monochrone Hercules or other cards, but they have not been extensively tested on such machines.

Color is used not only to distinguish different trajectories on the same graph, but also in maps to indicate the superimposition of plotted points.

The programs are designed for educational use only, and are not guaranteed in any way. The writer retains the copyright, but authorizes rewriting of the source code, and distribution of single copies of the programs, provided that the source is acknowledged and the programs are not sold for profit.

To start the programs, put the diskette in Drive `A:` (or `B:`) of your computer, change the default to that drive, and then type either `README` for more information, or `CHAOS` to begin the demonstration program.

If you require a paper copy of the graphs, use a *graphics print-screen* utility suitable for your own computer, or a *screen-grabber utility* such as one of those provided with many word processors.

The programs on the diskette include:

- `chaos` This program demonstrates several different dynamical systems, offered by the opening menu.

- `ranwalk` Demonstrates properties of random walks, including proper scaling of fractional Brownian motions, and diffusion by Brownian motion.

- `txtsplot` Plots a time series that has been saved as a `text` or `ascii` file.

- `tsplot` Plots a time series that has been saved as an `ascii` file, as a "scatter plot" to reconstruct the attractor using the method of delays (or derivatives).

- `crvsr` Analyses a time series using the Grassberger-Procaccia algorithm.

- **wimplex** The nonlinear prediction program published by Fowler and Roach (1993) modified to run on the PC.

Each of these is described in a separate section of this Appendix.

CHAOS

This program offers the following menu:

1. Logistic map;

2. Oscillator and forced Van der Pol oscillator;

3. Lorenz attractor;

4. Rössler attractor;

5. Sprott B attractor;

6. Hénon quadratic map;

7. Hénon attractor;

8. Circle map;

9. Quit.

Logistic map

This option demonstrates the logistic map, which is described in Chapter 2, and in most introductory texts on chaos.

Try running the program first with r in the range from 3 to 4 (using an arithmetic scale), and with x in the full range from 0 to 1. Next try r on a logarithmic scale in the range from 3 to 3.5699. This demonstrates the doubling of bifurcations with increments in r between bifurcations decreasing in a nearly geometric progression (equal spacing on the log scale).

Color is used to indicate the supimposition of plotted points. Each time a new point is plotted in the same position (i.e., pixel) as a previously plotted point, the color changes in the following sequence: 1. blue; 2. green; 3. cyan; 4. red; 5. magenta; 6. brown; 7.light grey; 8. dark grey;

9. light blue; 10. light green; 11. light cyan· 12. light red; 13. light magenta; 14. yellow; 15 white.

Oscillator and forced Van der Pol oscillator

This option is designed to introduce the concepts of phase space, trajectories in such spaces, and attractors. The example used is the oscillator:

$$\frac{d^2\theta}{dt^2} = k_1 f_1(\theta) + k_2 f_2(\theta)$$

where θ is an angular measure of displacement (the angle from the vertical, in the case of a pendulum). It is measured in radians and varies from $+\pi$ to $-\pi$. k_1 and k_2 are numerical coefficients, such as damping coefficients, and f_1 and f_2 are functions of θ.

The osillator is a second-order differential equation, so can be reduced to two first order equations by using the transformation $d\theta/dt = y$ as one of the equations. The phase space is two dimensional, and consists of the $\theta - d\theta/dt$ plane.

The program offers four options:

1. Simple pendulum.

$$k_1 = -g/L, \quad k_2 = 0, \quad f_1 = \theta$$

where g is the acceleration due to gravity, and L is the length of the pendulum. In the program k_1 is set equal to -1. Such a pendulum is an example of a linear Hamiltonian system: there is no friction, therefore no attractor. Trajectories in phase space are closed curves.

2. Nonlinear pendulum.

$$k_1 = -g/L, \quad k_2 = 0, \quad f_1 = \sin\theta$$

This is a nonlinear Hamiltonian system. The trajectories are very similar to those of the linear pendulum, but differ somewhat in shape.

3. Damped pendulum.

$$k_1 = -g/L,$$
$$k_2 < 0,$$
$$f_1 = \sin\theta,$$
$$f_2 = d\theta/dt$$

This system is dissipative (because k_2 is less than zero), and has a point attractor at the origin, which corresponds to the pendulum hanging vertically down. Trajectories spiral in towards the attractor.

4. Van der Pol oscillator.

$$k_1 = -g/L,$$
$$k_2 < 0,$$
$$f_1 = \sin\theta,$$
$$f_2 = (d\theta/dt)(\theta^2 - 1)$$

This system is also dissipative, but is driven by f_2. So, after an initial "transient", it settles into a regular orbit around the origin. The orbit is a one-dimensional (line) attractor, called a *limit cycle*.

A good introductory discussion of oscillators is in Bergé et al. (1986, Chapter 1), Baker and Gollub (1990, Chapter 3) and Rasband (1990, Chapters 1 and 6). Driven damped oscillators, with a third degree of freedom, can give rise to chaotic attractors of great complexity. An example is the *Duffing equation* (see Rasband, 1986, Chapter 6; Moon, 1992, Chapters 1 and 3).

Lorenz attractor

The Lorenz system of three ordinary differential equations is discussed in Chapter 2. The program is designed to display the equations in the range $24 < r < 28$, though that range can be extended somewhat: there are other interesting ranges of r, and the source code can easily be adapted to display them by adjusting the plotting scales.

In this and the Rössler and Sprott B options, the equations are solved numerically using a fourth order Runge-Kutta method (for a description see Press et al., 1989).

The program offers two choices:

1. A plot of the trajectory, either in the YZ or XZ projection, or in a "perspective" view;

2. A Poincaré section, for the plane $Z = (r - 1)$, and a first return map of $Z_{max}(k)$ against $Z_{max}(k + 1)$.

In both cases, time series of X, Y, and Z are plotted (in different colors) in a lower window.

The trajectory plots are useful aids in visualizing the geometry of the attractor. The Poincaré section appears to consist of two lines of dots, corresponding to the two "wings" of the Lorenz "butterfly." If the scale were enlarged, it would be seen that the "line" of dots is actually a narrow cloud. The fractal dimension of the attractor is only slightly larger than 2, so that of the section is slightly larger than 1.

The return map plots succesive maximum values of Z (i.e., one dot is plotted each time the trajectory goes over the top of the loop). This kind of plot was first presented by Lorenz himself: see the original paper or Bergé et al. (1986, Chapter 6) for a discussion of its significance.

Rössler attractor

This option is similar to that showing the Lorenz, except that it does not offer a Poicaré section.

Sprott B attractor

Sprott (1994) described 19 very simple sets of three ordinary differential equations that show chaotic behavior. This option displays his B attractor, which is somewhat similar to the Lorenz, but shows a more complicated attractor (even though the equations are structurally simpler). The more complicated attractor can be seen from the trajectories, and from the

Poincaré section and return map. The fractal dimension determined by Sprott (1994) is 2.17.

Hénon quadratic map

The program was modified from one published by Hughes (1986). It displays trajectories of the Hénon quadratic map—which is area-preserving and can be thought of as a Poincaré section through a non-dissipative (Hamiltonian) system. There is no attractor, and each trajectory starts from a different initial condition. Some trajectories are "closed" loops, some are a series of such loops (generally called "islands") and some give a chaotic scatter of points. See the entry **KAM Theorem** in the Chaossary (Appendix I).

Hénon attractor

This is a classic two dimensional mapping with a strange attractor. The program make it possible to zoom in on parts of the attractor, to see the increasingly detailed structure visible at higher magnifications. The attractor can be seen to be fractal in the sense that it is made of parts similar to the whole. According to Gleick (1987, p.149) Hénon devised the map because of the difficulty one of his colleagues encountered in seeing the fractal strcture of the Lorenz attractor. With the usual values for the parameters, the fractal dimension is about 1.2 (Grassberger, in Holden, 1986). A multifractal spectrum has been determined by Arneodo et al. (1987).

Circle map

The particular map displayed in a two-dimensional sine circle map. For further discussion see Procaccia (1988), Gunaratne et al. (1988), and Ott(1993, p.190–199).

The equations of the map are:

$$\theta_{n+1} = \theta_n + \Omega - (K/2\pi)\sin 2\pi\theta_n$$
$$+ Br_n \quad \text{mod } 1,$$

$$r_{n+1} = Br_n - (K/2\pi)\sin 2\pi\theta_n$$

These equations describe the motion of two oscillators. The frequencies are different, in the ratio Ω. Each oscillator will move independently of the other, unless they are coupled together in some way. K measures the strength of the coupling. B is a factor that controls the strength of the radial distortion: for $B = 0$, the orbits are true circles, for $B > 0$ the radius of the "circle" varies.

For $K > 0$, if the frequencies of the two oscillators are close to rational ratios (such as 1:1, 1:2, etc.) the phenomenon of *phase locking* occurs, i.e., after an initial transient, the frequencies adjust to the rational ratio, even if they were not exactly in the ratio to begin with.

To experiment with the circle map, chose a value of B (start with $B = 0$) and a value of Ω, and observe what happens as the value of K is increased.

RANWALK

This program demonstrates certain properties of random walks:

- Diffusion;

- Proper scaling;

- Fractional Brownian walks.

Diffusion is demonstrated by plotting 500 random walks produced by incrementing steps along the x-coordinate, and using a random number to determine the step in the y-direction. An "envelope" to the walks, with y_{max} proportional to the square root of x is also plotted. The y-coordinate after 200 steps is assigned to one of 20 y-bins. After all the walks have been simulated, the y-freqency histogram is plotted, a Normal curve fitted, and a goodness-of-fit Chi Squared value is calculated. In general, it will be found that there are significant differences from a Normal distribution—possibly because the walk has not been carried far enough.

Scaling of random walks is demonstrated by contrasting the appearence of the same walk at different horizontal scales, when the vertical scale is the same as the horizontal scale, and when the vertical scale is the square of the horizontal scale. The latter is demonstrated to be correct for Brownian motion. This part of the program is based on a program in Hastings and Sugihara (1994).

Fractional Brownian walks are produced (from the same random seed) by using the midpoint-displacement technique, using the algorithm given by Saupe in Peitgen and Saupe (1988).

TSPLOT

This program inputs a real time series, consisting of N values of $x(t)$, and plots it as a "scatter diagram" of:

1. $x(t)$ vs $x(t+T)$ for $t = 1 \ldots N-1$;

2. $x(t)$ vs $x(t+T)$ and $x(t+2T)$;

3. $x(t)$ vs dx/dt, defined as $x(t-1) - x(t+1)/2dt$;

4. $x(t)$ vs dx/dt and d^2x/dt^2.

The time delay, T, may be adjusted by specifying the lag.

TXTSPLOT

This program reads a time series stored as an `ascii` file, displays and plots the data on the screen, and calculates and plots the semivariogram—a form of autocorrelation function (e.g., see Davis, 1986, p.239–248).

CRVSR

This program is an implementation of the Grassberger and Procaccia (1983) algorithm for calculating the correlation dimension of a time series.

WIMPLEX

This is a modification of the program by Fowler and Roach (1993). The original program was written in Pascal for the Macintosh. It performs nonlinear prediction analysis, based on the suggestion by Sugihara and May (1990). Full details are given in the original articles—see also the discussion in Chapter 5 of these Notes. The program given here is a translation into Turbo Pascal. It uses pointers in order to conserve memory in the stack, and has no graphic output.

References

ARNEODO, A., GRASSEAU, H., and KOSTELICH, E.J., 1987, Fractal dimensions and $f(\alpha)$ spectrum of the Hénon attractor: Physics Letters A, v.124, p.426–432.

BERGÉ, P., POMEAU, Y., and VIDAL, C., 1986, Order within Chaos: Towards a Deterministic Approach to Turbulence: New York, John Wiley and Sons, translated from the 1984 French edition by L. Tucherman, 329 p.

DAVIS, J.C., 1986, Statistics and Data Analysis in Geology, Second Edition: New York, John Wiley and Sons, 646 p.

FOWLER, A.D., and ROACH, D.E., 1993, Dimensionality analysis of time-series data: nonlinear methods: Computers and Geoscience, v.19, p.41–52

GLEICK, J., 1987, Chaos: Making a New Science: New York, Viking,

GRASSBERGER, P., and PROCACCIA, I., 1983, Measing the strangeness of strange attractors: Physica D, v.9, p.189–208.

GUNARATNE, G.H., JENSEN, M.H., and PROCACCIA, I., 1988, Universal strange attractors on wrinkled tori: Nonlinearity, v.1, p.157–180.

HASTINGS, H.M., and SUGIHARA, G., 1993, Fractals: a User's Guide for the Natural Sciences: Oxford University Press, 320 p.

HOLDEN, A.V., ed., 1986, Chaos: Princeton Univ. Press, 324 p.

HUGHES, G., 1986, Henon mapping with Pascal: Byte, v.11, no.13 (December), p.161–178.

MOON, F.C., 1992, Chaotic and Fractal Dynamics: An Introduction for Applied Scientists and Engineers: New York, John Wiley and Sons, Inc., 508 p.

OTT, E., 1993, Chaos in Dynamical Systems: Cambridge Univ. Press, 385 p.

PEITGEN, H.-O., and SAUPE, D., eds., 1988, The Science of Fractal Images: New York, Springer-Verlag, 312 p.

PRESS, W.H., FLANNERY, B.P., TEUKOLSKY, S.A., and VETTERING, W.T., 1989, Numerical Recipes in Pascal: The Art of Scientific Computing: Cambridge Univ. Press, 759 p.

PROCACCIA, I., 1988, Universal properties of dynamically complex systems: the organization of chaos: Nature, v.333, p.616–623.

RASBAND, S.N., 1990, Chaotic Dynamics of Nonlinear Systems: New York, John Wiley and Sons, 230 p.

SPROTT, J.C., 1994, Some simple chaotic flows: Physical Review E, v.50, p.R647–R650.

SUGIHARA, G., and MAY, R.M., 1990, Nonlinear forecasting as a way of distinguishing chaos from measurement error in time series: Nature, v.344, p.734–741

Index

INDEX OF NAMES

INDEX OF TOPICS

An asterisk indicates an entry in the **Chaossary** (Appendix I). Note that a brief guide to mathematical notation may be found under **Symbols***.